from Sylvia Lo

IMAGES OF ASIA

Indonesian Batik

Other titles in the series

At the Chinese Table
T. C. LAI

Chinese Jade
JOAN HARTMAN-GOLDSMITH

A Garden of Eden: Plant Life in South-East Asia
WENDY VEEVERS-CARTER

The Kris: Mystic Weapon of the Malay World
EDWARD FREY

Macau
CESAR GUILLEN-NUÑEZ

Sailing Craft of Indonesia
ADRIAN HORRIDGE

Indonesian Batik

Processes, Patterns and Places

SYLVIA FRASER-LU

Warmest regards
Sylvia Fraser-Lu
Potomac, March 1994

SINGAPORE
OXFORD UNIVERSITY PRESS
OXFORD NEW YORK
1986

Oxford University Press

Oxford New York Toronto
Petaling Jaya Singapore Hong Kong Tokyo
Delhi Bombay Calcutta Madras Karachi
Nairobi Dar es Salaam Cape Town
Melbourne Auckland

and associates in
Beirut Berlin Ibadan Nicosia

OXFORD is a trademark of Oxford University Press

ISBN 0 19 582661 2

Printed in Singapore by Koon Wah Printing Pte. Ltd.
Published by Oxford University Press Pte. Ltd.,
Unit 221, Ubi Avenue 4, Singapore 1440

TO THE MEMORY OF
MY PARENTS
Archibald James Thomas Fraser
and
Dorothy Esther Fraser

Preface

WHILE the process of decorating woven cloth through the batik resist process is not unique to Indonesia, it has reached a degree of perfection in that country which is unequalled elsewhere. It is essentially an art developed by the people who inhabit the island of Java, which lies at the centre of a vast archipelago noted for its varied, intricate and labour-intensive textile traditions. The beauty of batik is a tribute to the patience, creativity and craftsmanship of the women of Java who are largely responsible for applying the batik decoration to the cloth. Credit too should be given to the men who prepare the cloth and handle the dyeing and finishing processes.

It has been estimated that there are over 3,000 batik designs in existence. Many of these reflect motifs which have been known in South-East Asia since neolithic times. Lying astride the trade routes between India and China, Indonesia has incorporated design elements from the very rich artistic registers of both these countries. The coming of the Europeans, Dutch colonization, Japanese occupation and post-independence nationalism have all exerted their influence on Indonesian batik.

Textiles and their designs have always been associated with traditions and festivals and are an indicator of status in Indonesia. Batik is no exception. Being strongly rooted in folk-craft as well as a royal tradition, batik designs and materials used vary from region to region, so adding to the infinite variety seen on this unique cloth.

This small publication aims to be but a brief guide or apértif for the general reader or traveller who has an interest in the process, designs and regional styles of batik as seen in Indonesia today. All the photographs in this book have been taken in

Indonesia and most batiks depicted, with the exception of those from the Kusuma collection, are generally still available for purchase today. For those who might wish to pursue the subject of batik in greater depth there are some useful general references in the 'Further Reading' section. Most of the books cited have excellent bibliographies for further consultation.

Bangkok, 1985 SYLVIA FRASER-LU

Contents

Acknowledgements

THE author would like to thank the following for their assistance while writing this book:

Mrs Pamela Reed, Mrs Rosel Hilliken, Mrs Elizabeth Mortlock and Ms Vipaporn Supasiri of Bangkok; Mrs Eiko Kusuma, Mr Iwan Tirta, Mrs Helen Cohen and Mrs Felicity Wood of Jakarta; Mrs Soerjanto, Mrs Sudevi and Mrs Hani of Jogjakarta; Ms Po Hwa Hwa and Mr Tio Swan Sien of Lasem; the Oey Soe Tjoen family of Kedungwuni; Mr Yahya of Pekalongan; the Masina family of Cirebon and Mr Robert Retka of Chevy Chase, Maryland, U.S.A.

1
Introduction

BATIK in Indonesia is made by decorating woven cloth with designs using a wax compound to cover those parts of the cloth which are to resist a particular colour during the dyeing process. The word *batik* is thought to derive from the Indonesian word *ambatik* meaning 'a cloth with little dots'. The suffix *tik* means 'little dot', 'drop' or 'point' but it can also denote a ticking or tapping sound. This root meaning may be seen also in words such as *tritik* (a Javanese word which describes a resist process by which designs are reserved on textiles by sewing and gathering before dyeing), *nitik* (batik designs which imitate weaving patterns), and *klitik* (the name of a well-known batik design). In a wider context, tik can be interpreted to refer to drawing, painting, and writing.

As with many traditional crafts, the origins of the batik process in Indonesia are obscure. The use of a resist (such as wax, starch, or any other dye-proof substance) to form patterns on textiles is at least 1,500 years old. Archaeological excavations in both Eygpt and the Middle East have brought to light fragments of batik-decorated cloth. Samples from Egypt date from the fifth and sixth centuries AD. East Turkestan, India, China, Japan, and West Africa also have established traditions of using the resist technique in textile decoration.

Some scholars believe that batik decoration came to Indonesia at a very early date from India, the origin of many diverse cultural influences on South-East Asia. Prior to the Industrial Revolution in Europe, Indian calicos, muslins, and silks were greatly sought-after trade items and formed the principal currency in the spice trade of the East Indies (Indonesia), being traded for cloves and cardamom. The presence of repetitive type textile patterns on the walls of ancient temples such as Prambanan (AD 800) and on stone statues dating to AD 1291 (Colour Plate 1) have led scholars to suggest very early dates for the beginning of batik in Indonesia. At this point, however, there is no conclusive evidence that the patterns depicted were produced by the batik process.

Such early origins for Indonesian batik have been challenged by other researchers who believe that the art of batik evolved fairly recently. As evidence they cite the fact that the word batik is not mentioned in the old Javanese language. There is also no mention of the batik process in the writings of fourteenth-century European travellers who were known to be acute observers of local phenomena. Batik is first mentioned in seventeenth-century Dutch sources in reference to a shipload of fabrics decorated with colourful patterns. Supporters of the recent origins of batik in Indonesia have stated that detailed Javanese designs were only possible on finely woven imported cloth, first from India until 1800 and after 1815 from Europe and (more recently) Japan. Local coarse-weaves were thought to be unsuitable for the intricate Javanese batik designs. Of still existing batiks, it is doubtful that any can be dated to earlier than the second half of the eighteenth century.

Some scholars suggest that *kain simbut*, a coarsely woven local cloth decorated with crude geometric and stick-figure designs which are applied using a resist of rice paste before being dyed red, could well be a precursor of Java's sophisticated wax-resist batik process. Kain simbut is still made in the more isolated areas around Banten in West Java. The Toraja peoples of Central Sulawesi also have a tradition of decorating material with geometric figures against a blue ground. The wax resist of the Toraja material is applied with a bamboo pen.

Evidence that earlier kain simbut or the Toraja cloth is a

precursor to Indonesian batik is not conclusive. It is clear, however, that both types of cloth were very important to their owners. Kain simbut was used in important rites of passage such as weddings and tooth-filing ceremonies, while the wearing of batik in Toraja society was restricted to successful head-hunters of noble families. Such restricted use of these materials may suggest long-standing traditional use. Scholars may disagree on the origins and age of batik making in Indonesia, but all agree that this uniquely Indonesian art form has reached a degree of perfection unequalled elsewhere.

It is commonly held that batik was originally the preserve of Indonesian royalty and aristocracy. It was the princesses and the noblewomen who had both the highly refined design sense and the time to draft the traditional patterns of consumate beauty and mystical importance. It is unlikely, however, that these women would have done more than the first waxing. The messy work of dyeing and subsequent waxings was probably left to court artisans who worked under aristocratic supervision. The feudal courts of Jogjakarta and Surakarta were known to reserve certain batik designs exclusively for royal attire.

Indonesian batik scholars such as Iwan Tirta and K.R.T. Hardjonogoro have suggested that there were two parallel traditions in batik, that of the palace and that of the *rakyat*, or common people. Small local batik industries still exist today in remote areas of north-east Java. Manual dexterity with the *canting* (the waxing implement) was formerly regarded as an important accomplishment for a young lady, being counted on a par with cookery and other housewifely arts.

Batik has become synonymous with Indonesia–the island of Java in particular. It is closely entwined with the other great art forms of Java, the *wayang kulit* shadow plays and the *gamelan* orchestras which accompany the *wayang* and classical dance performances. Not only did the Javanese *dalang* (puppeteer) of the wayang kulit preside over the most important of the performing arts, he was also an important source of batik patterns. When creating his puppets, he made perforated patterns of what they would wear. These were later sold in the village market to eager women who would copy the designs on batik by blowing charcoal through the holes.

Gamelan orchestras consist of gongs, xylophones, drums, flutes and string instruments resembling cellos or violas. These instruments are sometimes set on stands carved with motifs which are also used in batik. The names of well-known gamelan melodies such as 'Pisang Bali', 'Kawung', 'Limar', and 'Srikaton' have their counterparts in batik.

2

The Batik Process

Preparation of the Cloth

FINELY woven cotton and occasionally silk are used to make traditional batik. For the finest batik, the cotton must have a tight, evenly woven surface. Until the nineteenth century cotton cloth for batik came from such Indian cities as Madapolam and Calicut. It was first carried to Indonesia by Arab traders and later by the Portuguese, Dutch, and English. After the Industrial Revolution, Europe replaced India as the principal supplier of cotton cloth. Since the Second World War much of the cotton cloth for batik has been supplied by Japan. Today a number of Indonesian textile concerns make finely woven cotton cloth.

Four qualities of cotton are used in batik. The finest is called *primissima* and has a warp of 113 and a weft of 108 threads to 6.5 square centimetres. *Prima* is second in quality with a slightly lower thread count. Medium quality cloth with a thread count of 71 for the warp and 61 for the weft is called *biru* (blue). *Merah* (red) has a similar thread count but is considered a lower quality cloth.

Cloth for batik is traditionally measured in *kacu*, a unit based on the square. The specific length of a kacu is determined by the width of the cloth (usually 106 centimetres). The width of the cloth is folded back and forth diagonally across the length of the material with the size of the kacu being the length of each diagonal fold. After being cut into suitable lengths, the cloth is hemmed along the raw edges to prevent fraying.

The hemmed cloth is boiled and washed in water a number of times to remove the starch, chalk, lime, and other sizing elements present in the material. A few grains of caustic soda or potassium carbonate may be added to assist the process. To help the dyes penetrate the fibres, the cloth is soaked for a few days in a vegetable oil such as castor oil, or peanut oil. In former times the cloth would next be spread on a board and kneaded two or three times a day for six to twelve days. The cloth was thoroughly dried between each treatment. In preparing the cloth to accept special dyes, especially those made from natural dyes, the process could take up to forty days.

Traditionally excess oil was removed by boiling the cloth in an alkaline bath prepared from the ashes of rice stalks or banana trunks. Today a faster, cheaper, and more easily prepared soda compound is used. The material is then laid out to dry in the sun. To keep the threads in place the fabric is lightly starched with a solution made from boiled rice or cassava to which alum may be added. Lastly, the fabric must be ironed or pounded with a wooden mallet to render it smooth and supple and to prepare it to receive the wax design. With the finer machine-made cotton available today, pounding is generally omitted.

The preparation of the cloth is a task which is traditionally performed by men.

Waxing

Canting Method

The art of batik has reached its zenith in Indonesia largely due to the invention there of a unique implement–the *canting* (Plate 1). The canting is a small, thin-walled, spouted, copper vessel which resembles the bowl of a pipe. It averages about eleven centimetres in length when attached to a short bamboo or reed handle. The implement is filled with molten wax and held like a stylus. The batik artist draws designs on a length of cloth using the wax that flows from the canting's tiny,

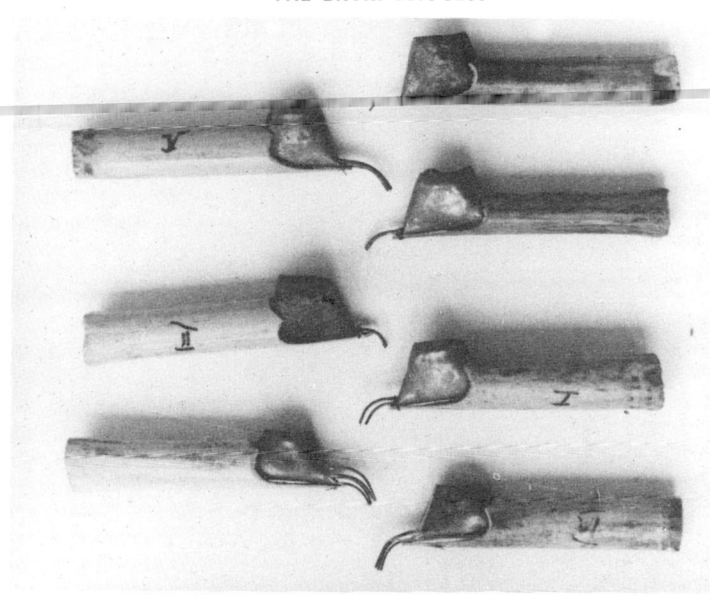

1. Canting showing spouts of various widths

downward-curving spouts. The number of spouts, their widths and endings, can be varied to achieve different effects with great precision. For fine detailed work a *canting rengrenang* with a spout about one millimetre in diameter is used. A wider spout is used to fill in larger design areas. Dots and parallel lines may also be drawn with canting having up to nine spouts. Some spouts may be arranged in a circle to enable the user to draw a number of dots at once. Very minute dots may be obtained by pricking the wax with a *cemplogen*, or comb of needles set in a wooden handle. A wad of cotton may be fastened over the mouth of the canting (or onto a stick) to serve as a brush to fill in extremely large areas.

The waxing process using the canting is performed by women who first lightly block out the main elements of the

2. Tracing a design from a stencil

design in charcoal or graphite. Designs may be traced from stencils (Plate 2), or patterns called *pola* may be purchased to be laid under the fabric and traced. Sometimes the pattern is laid out on a glass tracing table having a light beneath. More experienced artisans may bypass these steps and apply designs from memory directly to the cloth. Others copy from a sample.

Canting workers usually sit on mats or low stools in groups of four to six in front of material draped over light bamboo frames (*gawangan*) having horizontal rails (Plate 3). In the centre is an iron or earthenware basin called a *wajan*. The wajan is filled with molten wax and sits on a small brick charcoal stove or spirit burner called an *anglo*. The canting worker usually has a napkin (*taplak*) on her lap to protect her from the drippings of hot wax. Taking a piece of material in the palm

3. Applying the wax with a canting to cloth draped over gawangan

of her left hand, she dips the copper head of the canting into the molten wax. Grasping the implement with the tips of her thumb, first, and second fingers, she proceeds to draw designs on the fabric (Plate 4). The stem of the canting is held horizontally to prevent spillage of wax, for any accidental blots will lower the value of the finished piece. The canting should not touch the fabric while drawing the design. The left hand remains at the back of the fabric for support. An experienced canting worker is able to create the elegant cursive lines and intricate patterns for which Indonesian batik is famous. High quality batik decorated with the canting is referred to as *tulis* which literally means to 'write or draw'.

The wax consists of a mixture of beeswax (for malleability) and paraffin (for friability). Resins may be added for adhesiveness and animal fats for greater liquidity. There are two or

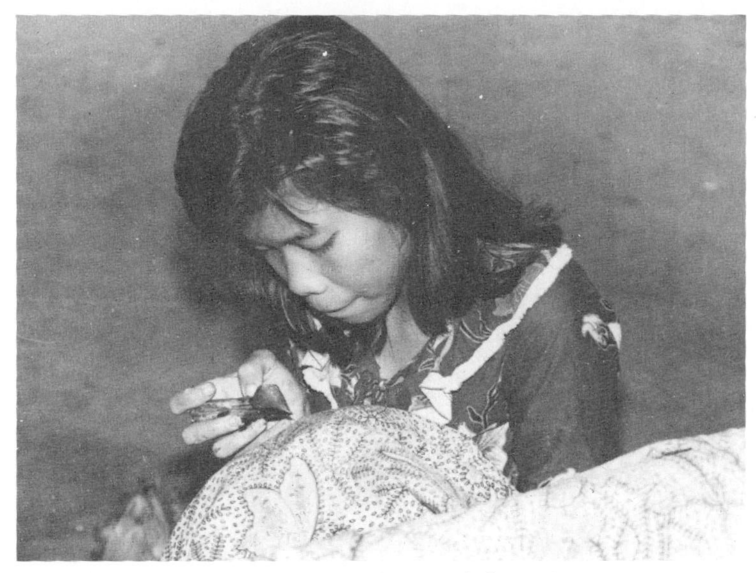

4. Applying the first waxing to a batik design with the canting

three types of beeswax from Timor, Sumbawa, and Sumatra and three types of petroleum-based paraffin–white, yellow and black–used in batik. The proportions of beeswax, paraffin, and additives are measured in grams and vary according to the design. For example, a higher proportion of paraffin may be added to increase brittleness if delicate tracery is an essential part of the design.

Some wax recipes are closely guarded secrets. Lighter and darker shades of wax make it possible to distinguish the various parts of the patterns in the course of the dyeing process. Outlines and detailed motifs are usually drawn with new light-brown wax to which a little old wax is added. The waxing of large areas is done with cheaper mixtures to which a higher proportion of resin may be added.

While in use, the wax is kept simmering on the stove. It is

strained through wire mesh from time to time to get rid of any impurities which may clog the canting. If too cool, the wax will clog the narrow spouts; if too hot, the wax will flow too quickly, be less controllable, and penetrate the surface of the cloth too deeply. The women workers are very good at regulating the flow of wax from the canting. The spout is kept clear of obstructions by blowing before the wax-filled canting is applied to the cloth. The spout may be cleaned more thoroughly from time to time by inserting coconut fibres.

To ensure that the pattern will be well defined, the waxing process is repeated on both sides of the cloth. True batik is reversible. For clarity of outline, the pattern must be matched exactly on the reverse side.

A division of labour is common, with the most experienced canting workers doing the first waxing. Subsequent waxing and the filling in of large areas may be entrusted to less experienced artists. Mistakes are very difficult to correct. If wax is accidentally spilt on the fabric, the worker attempts to remove it by sponging the affected area lightly with water. She then applies a heated rod of iron with a curved end to lift off the excess wax. Complete removal is virtually impossible. During the making of a piece of fine tulis cloth, over half of the time is spent on waxing with the canting.

Cap Method

Producing batik designs with the canting is an extremely time-consuming process. To meet the ever-increasing demand for batik in the nineteenth century, metal blocks, or *cap*, began to be used to apply the wax designs. With this implement, up to twenty lengths of cloth could be stamped in a day by a single worker.

The stamps consist of entire design units and are made of

thin, 1.5 centimetre-wide strips of copper set upright with smaller pieces of wire for dots. They are soldered to an open metal base of the same material. A curved iron handle is soldered to the reverse side (Plates 5 and 6). To make a cap requires great skill and precision on the part of the metal craftsman. With the aid of a compass and other drafting tools he transfers the design from a paper pattern onto metal strips. If two sides of the cloth are to be stamped, a design made of sheet copper may be welded between two metal, grid-like plates which form a base top and bottom. This block is cut in half at the centre parallel to the base so that the pattern on each half is identical. Dimensions and shapes in metal stamps vary, but for ease of handling, cap rarely exceed more than 24 centimetres in any direction.

To wax a cloth, several pairs of cap of different sizes and shapes may be used. As many as ten pairs may be needed for a complex pattern. In most cap work the pattern is stamped once, and various design areas are scraped away as the dyeing progresses. In some cases the process may be shortened by the *bedesan* method. In this process the material is restamped after the first dyeing with another design component which is to resist the second dye bath.

Cap designs are usually applied by men standing at square tables which are tightly covered with padded cloth or inset with flat, plastic-covered foam cushions (Plate 7). As with canting work, the molten wax is kept warm in a basin placed over a large spirit burner or on top of a circular concrete hearth built into the wall. Inside the basin is a wad of folded cloth approximately 30 centimetres square. Saturated with wax, it serves as a stamp pad. The cap is pressed onto the pad to absorb wax. It is then stamped onto the fabric to leave an imprint of the design in wax. The process is repeated until the cloth is completely covered. Design elements applied earlier may be protected from subsequent stamping by overlaying them

5. Cap depicting the garuda sawat design

6. Back view of the garuda cap

7. Applying a different cap design within a previously stamped border pattern

with paper replicas of the reserved design. On better quality cap work great care is taken to match the design exactly. Poor quality cap work may have overlapping secondary lines, lighter colours, and blank spaces where the design has not been carefully matched.

The designs on some fabrics may reflect the use of both the cap and the canting. The main outlines may be stamped with a cap, with the details being completed with the canting. In Indonesia both cap and canting types of work are done in most batik establishments.

In terms of quality and individuality, cap batik is generally disparaged by admirers of fine tulis work produced by using the canting. It should be remembered, however, that it was the cap which enabled batik to be produced quickly and economically. Without it, the Indonesian batik industry would most likely have suffered the same fate as other indigenous textile industries which had to compete with cheap mass-produced textiles from abroad.

Dyeing

After the initial waxing, the cloth is ready for the first dyeing. Traditional batik dyes came from a variety of natural and sometimes esoteric substances. These were concocted into complex recipes, the contents and quantities of which were jealously guarded and handed down from generation to generation of dye masters. Special rituals were sometimes performed and taboos observed to ensure a successful outcome, for a dyer's reputation formerly rested on his ability to produce cloth of a particular hue.

Dyeing was traditionally done in wide-mouthed, glazed earthenware receptacles. Today concrete vats of various sizes may be seen in batik establishments. Above these are racks and pulleys on which the fabric is hung during the dyeing process (Plate 8). Traditional cradle-like vats with wooden rollers are still in use in a number of establishments. Directly adjacent to the dyeing area is a drying space which may be an outside courtyard or an upstairs room. This area is strung with lines laden with brightly coloured fabric in all stages of the dyeing process (Plate 9).

The oldest and first dye to be applied in classical Indonesian batik was blue made from the leaves of the indigo plant (*Indigo tinctoria*), a one metre-high shrub native to South-East Asia. Indigo dyeing was usually done by men. One early twentieth-century account described the indigo dyeing process as follows: At midday on the day before the dyeing was to be done, the dyer mixed one gallon of indigo with five gallons of water. One-third quart of molasses sugar and an equal quantity of lime were added and the mixture was left to stand overnight.

8. Dyeing batik by passing the material under a roller at the base of a cradle-shaped vat

9. Batik cloth drying outside between dyeing

Through fermentation the sugar made the indigo soluble, while the lime made the solution alkaline. The process might be accelerated by adding the sap of the tingi tree (*Ceriops candolleana*) which acted as a fixing agent. Bits of chicken meat, a few drops of rain water, or a spoonful of kerosene or ashes from the kitchen might be included to propitiate any evil spirits. To ensure success, it was advised that domestic quarrels be avoided during the dyeing process.

In former times, light shades of blue would be produced by leaving the cloth in the dye bath for only a few hours. Deeper shades required repeated dipping (eight to ten times a day) for five or six days. Great care was taken to make sure that the cloth remained evenly submerged in the dye. On exposure to oxygen in the air, a strong blue colour emerged. Today the actual dyeing takes only a few minutes. For blue, naphtol and indigosol chemical dyes are used.

On traditional batik *soga*, a warm brown colour ranging from a light yellowish tan to a deep chocolate brown, was usually applied on completion of the indigo dyeing. This natural dye, which is still quite widely used today, is made by soaking the bark of the soga tree (*Pelthophorium ferrugineum*) in a lukewarm bath. Another natural dye still in use today is a deep turkey red called *mengkudu* produced from the bark and leaves of *Morinda citrifolia*. Natural yellow tones may be produced from the bark of the jirak tree *Symploco fasciculata* and *Sophora japonica*. When used these vegetable dyes are combined with local tree resins. The depth of colour depends on the number and frequency of immersions over a given period of time.

Today, apart from the three colours mentioned, imported dyes, which are easier to prepare and handle, have generally replaced natural dyes. There are numerous manuals in the Indonesian language which give detailed instructions for mixing chemical dyes. The steeping time has been reduced from days to hours and minutes. In many cases only dipping is required. Despite this simplification, the dyer must be able to blend his chemicals skilfully, possess a steady hand, and have an astute sense of timing. Chemical dyes produce more predictable results and offer better resistance to sunlight and frequent washing. The great disadvantage of vegetable dyes is that they fade in strong sunlight. Greater uniformity in the dyeing process, however, means that regional batik styles are no longer distinct. In former times an expert could tell the provenance of a batik by the shades of colour, the motifs, and the tulis work.

The traditional batik dyeing process began with a waxing of all parts of the design that were to resist the blue dye. This included areas which were to be left white or dyed another colour at a later stage. The fabric was immersed in a vat filled with indigo dye and then soaked in cold water to harden the wax.

Once the blue dyeing was completed to the dyer's satisfaction, wax from the areas to be dyed brown was scraped away with a small knife. These areas were sponged with hot water and the cloth was resized with rice starch. The areas which were to resist the brown dye (blue and white) were covered with wax. Additional wax was scraped away or reapplied for other colours. If a marbled effect was desired, the waxed surface was cracked and redipped in the dye. It is the seepage of dye into the cracks which creates the fine veins of colour associated with batik. Traditionally cracks were not desired particularly where a blue colour was concerned, and were regarded as a sign of inferior craftsmanship. On brown pieces, however, cracks were permitted.

With skilled dyeing by traditional means many colour nuances were possible. These included light and dark blue, green (from blue and yellow), and purple (from blue and red). Soga brown combined with indigo blue gave black.

Today, if more than two or three colours are required,

minor areas of colour may be painted directly onto the cloth with a brush and then waxed for protection during the dyeing process. Direct painting on cloth is referred to as *colet*.

Once all the colours are obtained, the material is washed and soaked in a lime solution followed by a fixing bath of local sugar, alum, and whiting. Finally, the material is thoroughly rinsed and transferred to a concrete cauldron of boiling water. Here the wax is melted off and saved for future use. A small residue of wax, however, tends to linger in the fabric and gives new batik its characteristic stiffness and smell.

It can be seen that producing designs in batik is a lengthy process. The decoration of good quality cloth may take anywhere from five weeks to over a year.

Prada or Gold-decorated Batik

For festive occasions and ceremonial use some batik was formerly highlighted with gold leaf or gold dust on one side or on half of the cloth. Such gold-decorated batik is known as *prada* cloth.

Chinese gold leaf was applied to cloth with albumen in Jogjakarta and Surakarta. On the north coast of Central Java, fabrics were decorated with gold dust instead of gold leaf. The gold was applied to the fabric with a glue made of egg white or linseed oil and yellow earth. The gold remained even after laundering.

The gilding of the cloth could follow the underlying batik design or form an independent design. Old festive cloths were sometimes given a new lease on life by being decorated with gold. A wooded landscape design rendered in cream, indigo, and soga brown was very popular in prada work made in Central Java. Symmetrical designs were generally preferred on prada cloth destined for Bali. Gold-decorated batik cloth continues to be made today. Gold paint rather than gold leaf is generally used.

3
Traditional Batik Clothing

INDONESIANS have generally used batik to decorate items of clothing. Due to a warm tropical climate, traditional Indonesian clothing is fairly simple. It consists of untailored rectangular pieces of cloth which rely on folding, wrapping, and draping for a comfortable fit.

Kain Sarong or Tubular Cloth

A most important garment since Dutch colonial times has been the *kain sarong* which covers the body from the waist to the ankles. It was formerly only worn by women, but it is now also worn by men for relaxing at home or for going to the mosque. It consists of a piece of cloth approximately 215 centimetres long by 106 centimetres wide. This rectangle of cloth is sewn at the sides into a tubular skirt. When worn, it is stepped into, pulled up, and wrapped around the body before being drawn tight on one side. For women the excess cloth is usually folded over into a single fold or gathered into a number of small pleats which may be secured by a band of cloth, a belt, or merely tucked in at the waist. For a man the excess fullness of cloth is gathered in front and is kept in place with either a belt or a knot.

Traditional sarongs for women have narrow borders along the sides of the cloth. These may be undulating or straight. There is a central panel called the *kepala*, or head, which differs in pattern and colour from the main field, or *badan* (Figure 1). The kepala consists of a vertical border, or *papan*, with two rows of facing isosceles triangles called *tumpal*. The tumpal design may sometimes appear at each end of the cloth. The

Figure 1. Sarong with a kepala featuring a tumpal design between a floral badan

tumpal section of a sarong is usually worn in the front by women and at the back by men. On some cloth, the kepala may be delineated by a contrasting colour zone rather than triangles. Apart from colour, the motifs in each zone may be identical or they may be of a completely different design.

Kain Panjang or Long Cloth

A longer version of the sarong is the *kain panjang*, or long cloth, which is more than twice as long as it is wide (approximately 250 centimetres long by 106 centimetres wide). It is a skirt cloth worn by both men and women. It is usually decorated with an all-over pattern in a traditional batik design. This garment is worn draped around the waist. The last section is usually arranged in a series of pleats in front which fan out when the wearer walks. This garment may be referred to as a *bebed* for men and a *tapih* for women. While the sarong may be regarded as everyday wear for men and women, the kain panjang is usually worn on more formal occasions. It was formerly a very typical garment of the courts of Surakarta and Jogjakarta.

Figure 2. Kain panjang featuring two distinct pagi-sore design areas with tumpal at both ends

Figure 3. Kain dua negri: the kepala shows a floral bouquet typical of Pekalongan while the badan is in a traditional Central Javanese parang design

Sometimes a kain panjang may feature a different design on each half of the cloth. The division may be vertical or diagonal. This contrasting design is referred to as *pagi-sore*, or 'morning and evening' cloth (Figure 2). By reversing the folding, the owner is able to make use of two different designs from a single piece of cloth. In fact, the wearer can use the darker design for day wear and the lighter one for evening. On the best quality work, skilful changes in background design rather than in the overlying motif render a most pleasing contrast.

Some batik cloth has been referred to as *kain dua negri*, or 'cloth of two countries' (Figure 3). One section of the cloth (usually the pagi-sore type) may be first decorated with a batik design in Central Java. The partially completed cloth may then be sent to the north coast of Java for completion by another artisan. For example, the main field, or badan, of a cloth might be decorated with a traditional repetitive design in Jogjakarta and sent to Pekalongan to have the kepala embellished with a floral design.

Dodot or Ceremonial Cloth

Even longer than the kain panjang is the *dodot*, a ceremonial cloth formerly worn by members of royalty, the aristocracy, and court dancers (Figure 4). It may also be worn by a bride and groom on their wedding day. It consists of two lengths of cloth sewn together and may measure anywhere from 2.10 to 4.6 metres in length. It is approximately six times the length of an ordinary sarong. The dodot is usually patterned with an all-over design. Some may have a lozenge-shaped, plain centre field called a *tengahan*. The dodot formerly was worn draped around the body in a variety of ways depending on court etiquette. It might have a train or a large overhanging fold on one side. Men wore the dodot at knee length with tie-dyed silk trousers underneath. Women usually wore the dodot as a long dress pulled tight across the breasts in a strapless style with flowing drapery down the front or back. Occasionally the kain panjang might be worn as a skirt underneath the dodot.

Kemben or Breast Cloth

Indonesian ladies formerly wore a *kemben*, or breast cloth, a long narrow strip of material which, when wound tightly around the chest, left the shoulders bare (Cover Plate).

Figure 4. Dodot

Measuring about 75 centimetres in width by 250 centimetres in length, it was sometimes decorated with an elongated plain coloured lozenge shape called a *blumbangan* or a long narrow oval called a *sunangan* in the centre. This centre lozenge was surrounded by small stylized leaves called *cemukiran*. For extra strength, this plain centre area might be lined with silk underneath. The kemben today has largely been replaced by the more modest but tight-fitting tailored blouse called *kebaya*.

Selendang or Shawl

Similar in size and shape to the kemben is the *selendang*, a scarf or shawl, which can be draped around the head or upper torso in a variety of ways. It may also be tied into a sling and used as a carry-all for babies, baskets of food, and personal objects (Colour Plate 2). It sometimes comes with a matching kain panjang. Court dancers wear a longer version of the selendang called the *sonder*, a long scarf tied around the hips with the ends hanging to the floor. The skill with which dancers manipulate the sonder is an important element in Javanese dance. Selendang are sometimes made of silk and usually terminate in a tumpal border or in a series of vertical lines along the selvages. The narrowest edges may be finished with silk and cotton fringes.

Iket or Head-cloth

For formal occasions men may wear a starched head-cloth, or *iket*, measuring about 85 square centimetres (Colour Plate 3). The iket usually has an all-over traditional design. Like the kemben, the iket design may be relieved by a plain square or lozenge-shaped central area, the tengahan. This tengahan may be tie-dyed and is usually bordered by stylized cemukiran leaves (Figure 5). Right-angled border patterns may be filled

Figure 5. Iket in a kawung pattern bordered by cemukiran leaves with a diamond-shaped tengahan

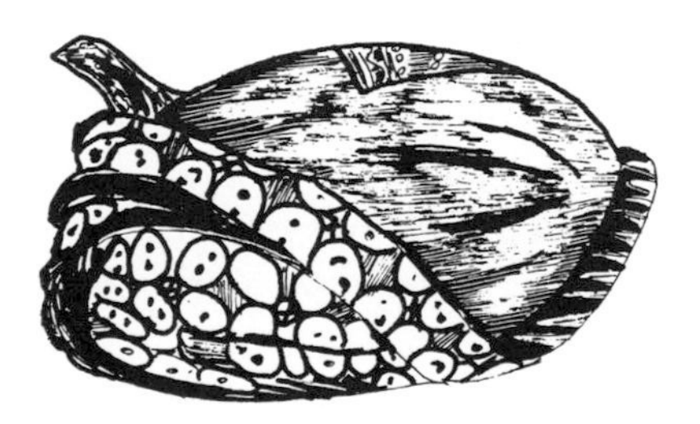

Figure 6. Iket folded in the form of a turban

with a butterfly or flower motif. The iket head-cloth can be tied in a variety of ways to form a turban (Figure 6). It was at one time possible to tell the rank and provenance of a man in Indonesia by the way he wore his turban. Today iket are generally purchased ready-made into a cap of stiffened material.

Pants

During the late nineteenth and early twentieth centuries colourful loose-fitting pants were made of batik. They were worn by Chinese and Dutchmen for relaxing at home.

Other Uses

Altar Cloths

Altar cloths and banners for Chinese temples were traditionally embroidered on silk. Since these were not readily obtainable in Indonesia, coupled with the fact that many Chinese were involved in the batik industry, altar cloths came to be made of cotton and were decorated in Chinese designs using the batik process (Colour Plate 4).

Wayang Golek Puppets

Javanese *wayang golek* stick puppets are always clothed in skirts featuring traditional batik patterns (Colour Plate 5).

4
Batik Designs

IT has been estimated that there are over 3,000 batik designs in existence. Many of these reflect indigenous motifs rooted in neolithic culture; however, over the years, designs from Indian cottons, Chinese textiles, ceramics, carvings, and European floral patterns have been added to the Indonesian batik repertoire.

In Indonesia, textiles have always been associated with traditions, festivals, and religious ceremonies. Rites of passage, particularly on the outer islands of the archipelago, have always involved specially woven and decorated cloths. Indonesians have generally been firm believers in the mystical qualities of certain textiles and designs. Some patterns have been credited with the ability to ward off ill fortune, others bring good luck to the wearer, while some are believed to help ensure fertility. For this reason, certain batik designs are worn at weddings by the bride, groom, and their respective families to ensure an auspicious day and a fruitful union.

Textiles and their designs often indicated the status and origin of the wearer. Certain batik designs were reserved for the rulers and their families. The rank of a prince could readily be determined by the batik designs he was allowed to wear. Other textiles served to appease the gods. The Goddess of the South Seas, Ratu Kidu (formerly regarded as a formidable deity), was regularly placated with offerings of textiles.

Batik designs are visually very pleasing. They reveal a taste for rhythm and complexity in both geometric and freehand designs. The best in batik design shows a distinct harmony between the broad patches of colour and the line of the principal design and the subtle detailed filling of the background pattern.

Isen or Background Designs

Isen patterns consist of simple repetitive design elements based on dots, lines, squares, crosses, foliage, and flowers (Figure 7). On some of the plainer batik, an isen motif might be the sole design element.

One of the oldest of the isen designs is the *gringsing*, or fish-scale motif (Colour Plate 6). This design consists of a series of nucleated circles or semicircles placed closely together. Gringsing was at one time considered effective against sickness, so it was worn in the hope of warding off illness. Other circular isen include patterns with names such as rice grain (*upan*), pigeon's eyes (*mata dara*), buttonholes (*uter*), shining scales (*sisik*), and the seven dots design (*cecek pitu*).

Designs based on straight, undulating, parallel, and diagonal lines include those with names such as the chequer-board (*poleng*), petal veins (*cecek sawat*), river fish (*uceng*), and roof-tile (*sirapan*). More flowing designs are based on lozenge shapes (*mlinjon*), rice stalks (*ada ada*), coconut fronds (*blarak sahirit*), tendrils of plants (*cantel*), and hooks (*ukel*). Swastika, or *banji*, designs (Figure 8; Colour Plate 7), emblems of good fortune, form very arresting isen patterns by providing a grid-like frame or background in direct contrast to the floral patterns which are often found on the same length of cloth. Other well-known Chinese lattice patterns such as the key, thunder, spiral, and 'T' may also be seen on batik. Chequer-board squares with dots in the centre, cruciform motifs, and other designs with intersecting lines are also popular filler patterns, as are zigzags and lozenges aligned in a variety of ways.

Stylized representations of well-known flora of Indonesia such as rice stalks, coconut fronds, curling leaves, buds, tendrils, and simple floral motifs are also represented in isen patterns.

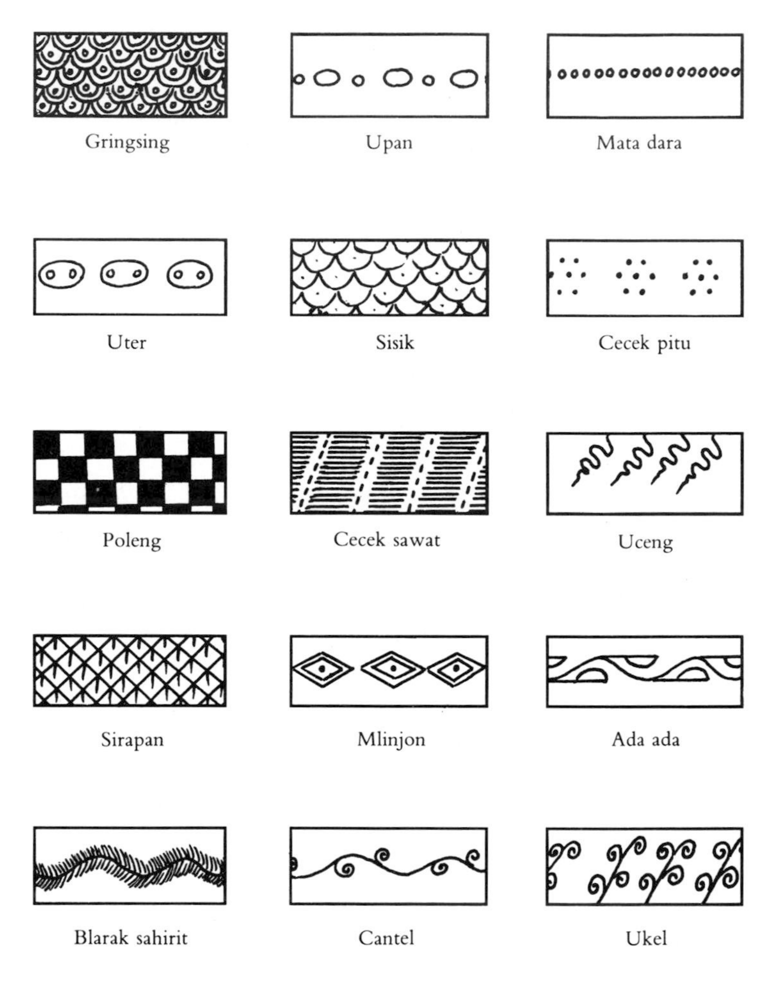

Figure 7. Isen

Banji Banji kasut

Figure 8. Banji

Geometric Designs

Ceplokan or Repetitive Designs

Ceplokan designs consist of symmetrical motifs in the form of stars, crosses, rosettes, lozenges, or polygons as seen from above (Figure 9). Some may be extremely simple consisting of a single motif or pair of motifs (such as the *grompol* or *truntum* (Colour Plate 8)) spaced at regular intervals over the surface of the cloth. Conversely, others may be models of complexity in which a number of design elements coalesce into different patterns. The entire design may be contained within a circle or square in a symmetrical arrangement and be aligned on a horizontal or diagonal axis. Ceplokan designs can be very striking. Variations in colour intensity create illusions of depth, and the overall effect is not unlike medallion patterns seen on Turkish tribal rugs.

In ceplokan the forms of nature such as flowers, fruit, birds, insects, and animals are portrayed in a very conventionalized way. The Indonesian population is largely Muslim, a religion that forbids the portrayal of animal and human forms in a realistic manner. To get around this prohibition, the batik worker does not attempt to express subject matter such as flora and fauna in a naturalistic way. Rather, the artist takes

Grompol, a cluster

Surketan, a grass design

Swelogiri, a floral design

Supit urang, pinchers of the lobster

A geometric design

A gangong motif within a woven pattern

Figure 9. Ceplokan

one or more of the key elements in a plant or an animal (such as a bud, a leaf, a seed, a feather, or a tail) and through elaborate embellishment, constructs it into a distinct design element. At first, the source of these motifs was fairly recognizable. With the evolution of the motif in the hands of new generations of batik artists, the original inspiration for the design is no longer always evident. In many cases, the name of the design element continues to provide a clue.

Indonesia's well-known flowers and fruit such as the hibiscus, jasmine, lotus, cotton plant, banana, mangosteen, and *salak* fruit have all been portrayed in stylized form on batik. Various spices such as cloves, nutmeg, betel nut, and the coconut palm have all been named in batik design. Elements of *gangong*, a leafy plant with fibrous flowers, are very popular in ceplokan designs. The *kusomo* plant, or 'flower of victory', is a popular motif in batik. This flower has come to represent longevity and a lasting reign.

Members of the animal world are also seen in conventionalized designs. Butterflies are a popular motif, as are water creatures such as *bibi* mussels, fish, prawns, and the claws of the crab. Natural phenomena such as moonbeams, stars, and swirling water may also be expressed in ceplokan motifs. Abstract concepts in the realm of beliefs and feelings have also lent their names to batik designs. For example, it is not unusual to come across batik designs called 'joy of meeting', 'pining for a loved one', 'defender of the faith' and so on.

One very important ceplokan design is the *sido-mukti* pattern (Colour Plate 9) which consists of small motifs such as pavilions, garuda wings, plant tendrils, butterflies, and other insects enclosed within bands of lozenges aligned diagonally across the material. The background may be plain or it may be covered with a fine isen in the form of the *ukel*, or hook pattern. Sometimes a plain background alternates with the ukel pattern to give a chequer-board appearance to the cloth. This design is associated with a glorious and untroubled life and may be worn by a bridal couple on their wedding night.

Ceplokan designs have been strongly influenced by the Gujerat silk textile called *patola* which is decorated with repetitive floral and geometric patterns produced by the double *ikat* process of dyeing designs on the wrap and weft yarn prior to weaving. This particular type of imported textile was highly prized in Indonesia and in some places was part of bridal attire. In Bali, patola was used for temple hangings while on the island of Sumba, it was reserved for the aristocracy. Because of their popularity, patola designs were imitated and modified, eventually becoming an integral component of the Indonesian design inventory.

Kawung or Circular Designs

One of the oldest and most famous of the ceplokan designs is the *kawung* which consists of parallel rows of ellipses (Figure 10). These ellipses may be embellished inside with two or more small crosses or other ornaments such as intersecting lines or dots (Colour Plate 10). It has been suggested that the ovals might represent flora such as the fruit of the *kapok* tree or the *aren* (sugar palm).

The kawung design has been known in Indonesia at least since the thirteenth century. During the Majapahit period (AD 1292–1389) it appeared on stone sculpture from Kediri in East Java and on the walls of the Sivaite temples of Prambanan near Jogjakarta. This design was also known in the early civilizations of Crete, Northern Syria, East Persia, and the Indus valley. It is a favourite motif of Central Java.

There are many variations of the kawung pattern. Instead of the usual diagonal alignment, the circles may be aligned on a square. They may be widely spaced or appear almost to coalesce. Sometimes the circles are shaped like flower petals or are flattened like sections of a fruit. The kawung design may sometimes provide a background for other batik motifs, or it may be alternated with some other design. On the more elongated, widely-spaced examples, the central lozenge assumes similar proportions to the ellipses and portrays an additional design element. Sometimes the cross rays of the ellipses appear to intersect and overlap to form the *kawung*

Kawung sari — Kawung sawo

Kawung dudo nggamblok — Kawung kembang

Kawung picus — Kawung ageng

Figure 10. Kawung

picus (coin) design which is based on the ten cent coin (Colour Plate 11).

Jelamprang or Designs Based on Indian Textiles

Closely related to the kawung is the *jelamprang* design, a repetitive, eight-rayed rosette motif set in squares, circles or lozenges, the boundaries of which touch but do not overlap. This design is said to symbolize Allah's nine *walis*, or defenders of the faith. The design also draws its inspiration from the patola motif and may be seen on Indonesian ikat work. Some of the circular jelamprang patterns are referred to as *cakar* and perhaps represent the *dharma* or 'wheel of the law', an important concept in Buddhism.

Nitik or Weaving Designs

Many of the jelamprang designs are composed of small dots and lines imitative of weaving designs called *nitik* (Figure 11). Stars, squares, crosses, and circles are similar in composition to ceplokan designs and are usually portrayed on a plain dark ground. Some of the simplest nitik designs such as *nam tikar* ('woven bamboo') imitate mat and basket weave patterns and

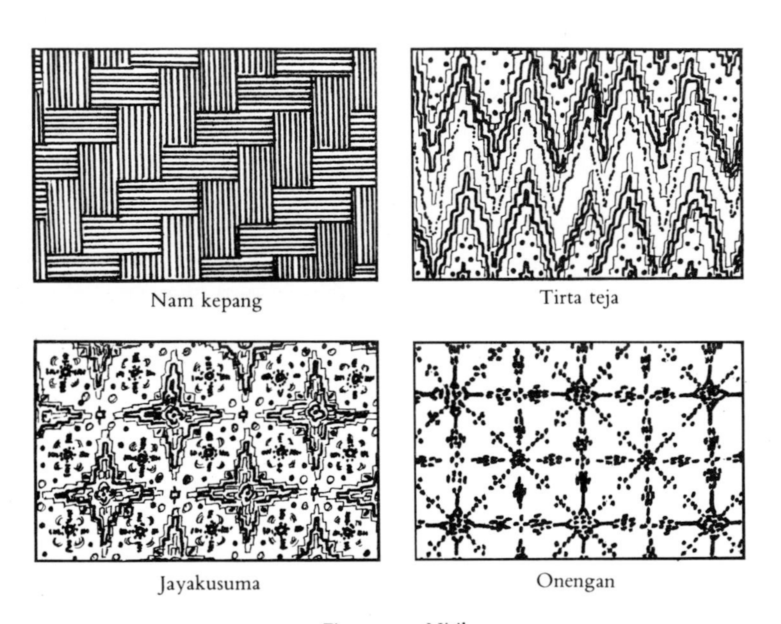

Figure 11. Nitik

are among the oldest batik designs. One of the most important of the nitik patterns is *tirta teja* ('bright water'), a pleasing design in the form of horizontal zigzags. Others have names such as *jayakusuma* ('flower of victory') or *onengan* ('to pine for a loved one').

Garis Miring or Parallel Diagonal Designs

Diagonally aligned designs in Javanese batik are referred to as *garis miring*. They are among the most visually striking of all batik patterns. They have a pronounced slimming effect on the wearer and are considered to be fortunate patterns.

PARANG OR KNIFE DESIGN

The most famous of the garis miring (diagonal designs) is the *parang* which has several related meanings such as 'rugged rock', 'knife pattern', or 'broken blade' (Figure 12). It consists of a series of broad light-coloured bands bounded by undulating or scalloped edges. The parang usually alternates with narrower bands in a darker contrasting colour. These darker bands contain another design element in the form of a line of lozenge-shaped motifs called *mlinjon*.

The parang motif appears in artistic media other than batik, being seen in wood carving and on Indonesian gamelan instruments. This design was traditionally reserved for the robes of the sultan's family and formed part of the offerings to the spirits of the royal ancestors. It was also offered to appease the Goddess of the South Seas.

Over forty parang designs are known. The most famous is the *parang rusak* which in its most classical form, consists of rows of softly folded, slightly bloated undulating ribbons of creamish white aligned between parallel rows of large dots of the same colour (Colour Plate 12). These rows may be separated by narrow concave bands of a dark blue-black

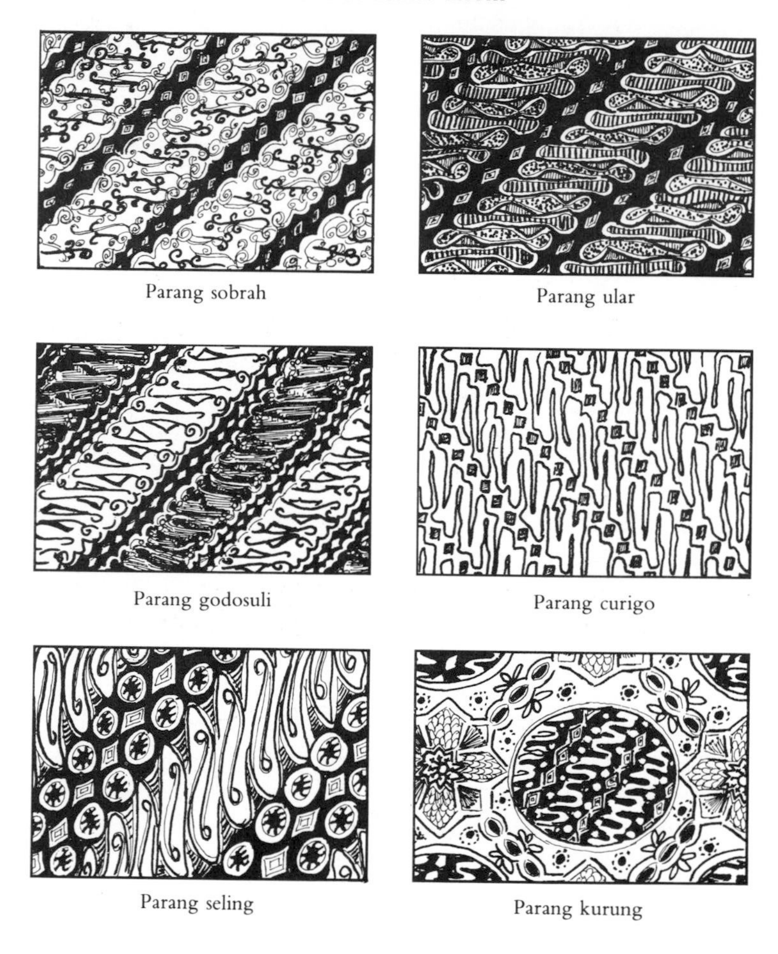

Parang sobrah

Parang ular

Parang godosuli

Parang curigo

Parang seling

Parang kurung

Figure 12. Garis miring: parang

colour set at regular intervals with mlinjon in soga brown. The largest of the parang rusak designs is called *barong*. A much smaller example is referred to as *parang klitik*.

Some parang patterns have line ornaments radiating from

1. Harihara portrait figure of King Kertajasa, Candi Sumberjati, East Java, *c.*1294–1309, wearing finely patterned garments decorated in ceplokan and kawung designs. National Museum, Jakarta

2. Women of Bali dressed in the traditional sarong kebaya

3. A custodian at the kraton of the Sultan of Jogjakarta wearing a kain panjang and an iket

4. Batik altar cloth formerly from a Chinese temple. Probably made in Cirebon *c.*1920. Collection of Mr and Mrs Hilliken, Bangkok

5. Wayang golek puppet figure of Gatokaca from the Mahabharata epic in a skirt of parang patterned batik

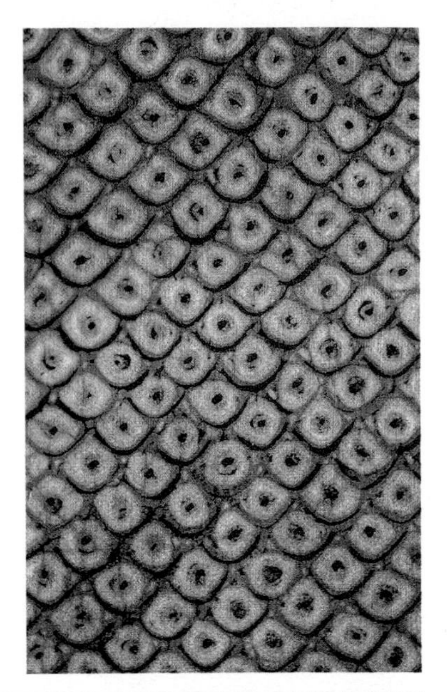

6. Gringsing design

7. Banji design. Cirebon, 70–80 years old. Collection of Mrs Kusuma

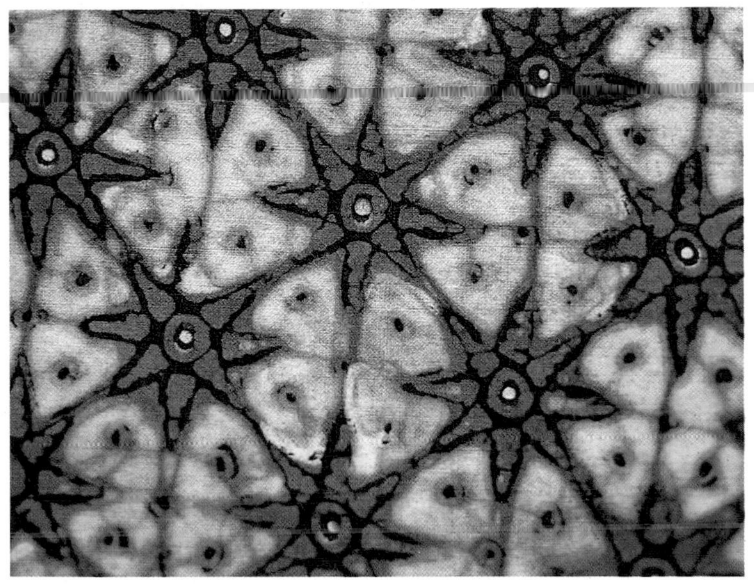

8. Truntum design, tulis, Iwan Tirta, Jakarta

9. Sido-mukti design from Surakarta, tulis, Iwan Tirta, Jakarta

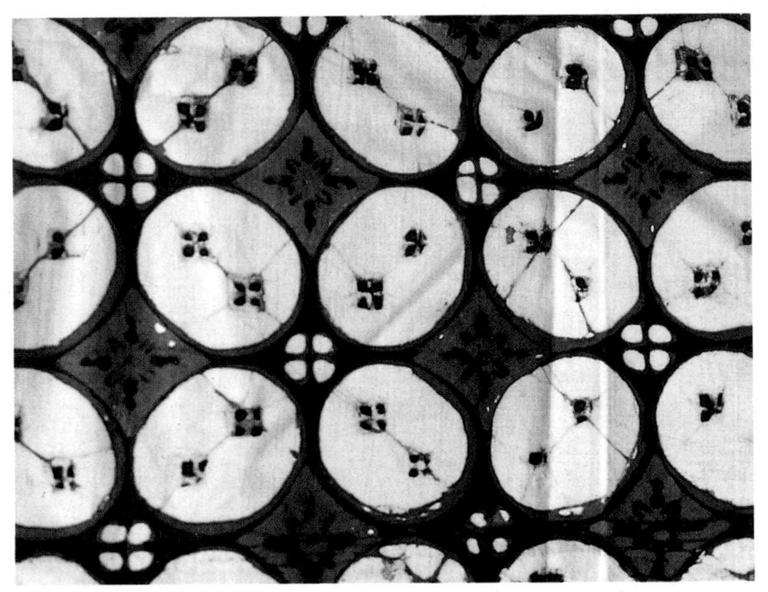

10. Kawung design, cap

11. Kawung picus design, cap

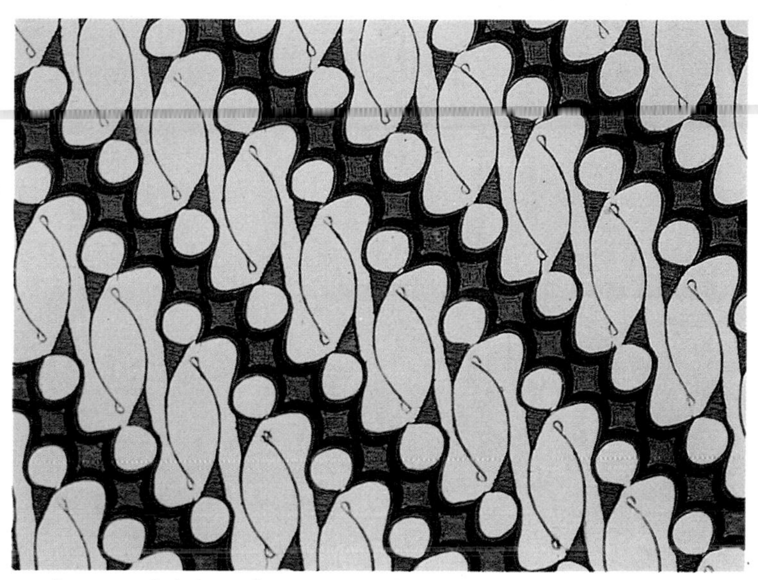

12. Parang rusak design, tulis

13. Parang baris design. Probably made by K. R. T. Hardjonogoro, Surakarta. Collection of Mr and Mrs Hilliken, Bangkok

14. Udan liris design, tulis, Hadiwasito Workshop, Jogjakarta

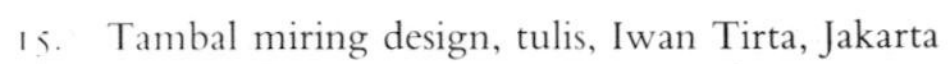

15. Tambal miring design, tulis, Iwan Tirta, Jakarta

16. Kumudaretna design embellished with gold leaf, tulis. Collection of Iwan Tirta

17. Garuda wings and tail in the sawat design against an udan liris background, cap

18. Fanciful lion motif, tulis, Iwan Tirta, Jakarta

19. Alas-alasan design, Central Java, *c.*1900

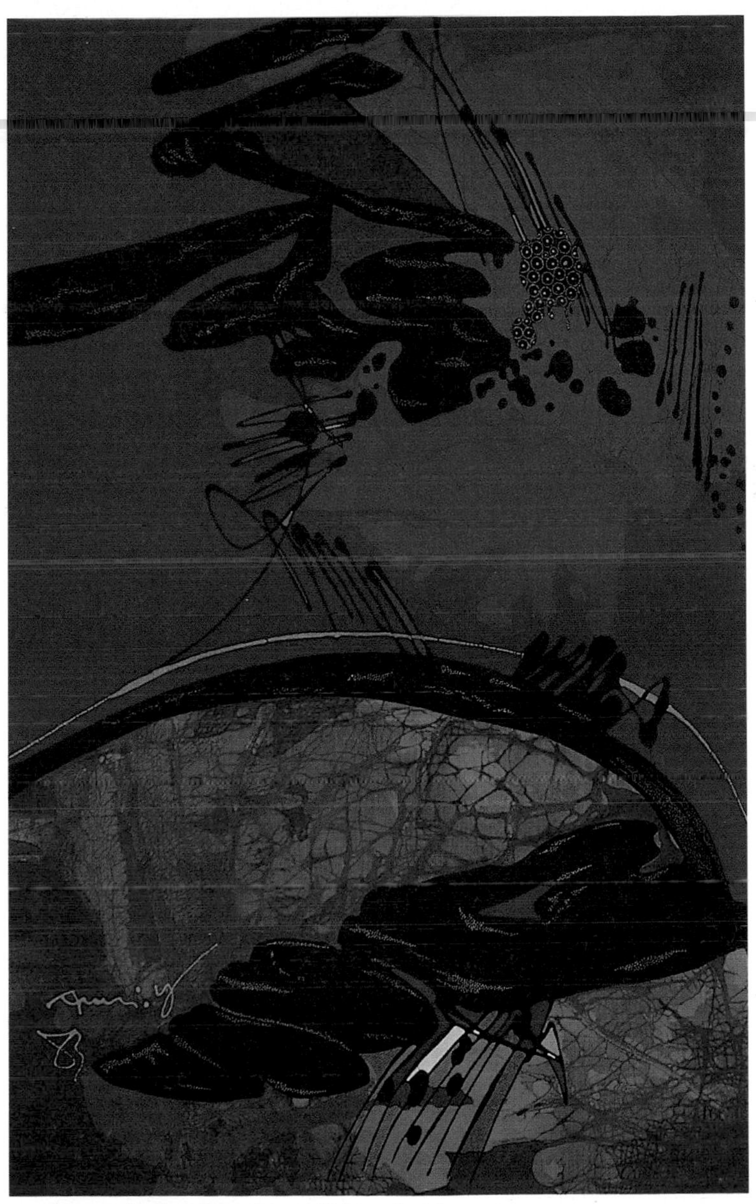

20. Abstract batik painting, Amri Yahaya, Jogjakarta

21. Solo malam design, 65 years old. Collection of Mrs Kusuma

22. Part of a shirt length, tulis, Obay Hobart Workshop, Tasikmalaya

23. Dutch inspired flora bouquets against a plain background, tulis, Eliza Van Zuylen, Pekalongan, *c.*1920. Collection of Mrs Kusuma

24. Bunga sepatu, Hokukai batik, tulis, M. Alhamani, Pekalongan, 40 years old. Collection of Mrs Kusuma

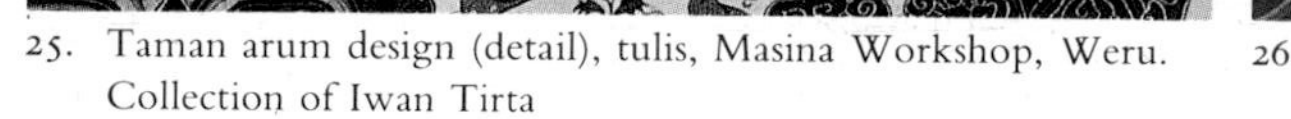

25. Taman arum design (detail), tulis, Masina Workshop, Weru. Collection of Iwan Tirta

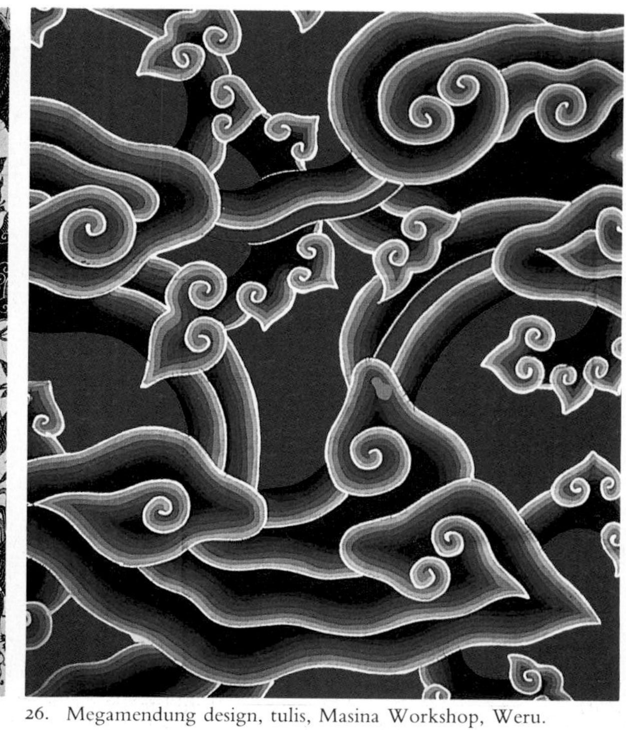

26. Megamendung design, tulis, Masina Workshop, Weru. Collection of Dr and Mrs G. V. Wood, Jakarta

27. Tiga negri design, tulis, Van Chellem Workshop, Lasem

28. Selendang featuring Cirebon inspired lokcan, tulis, Kerek

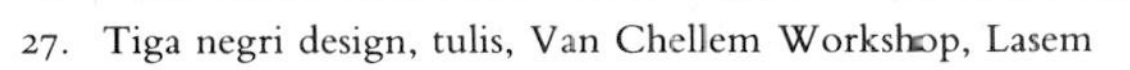

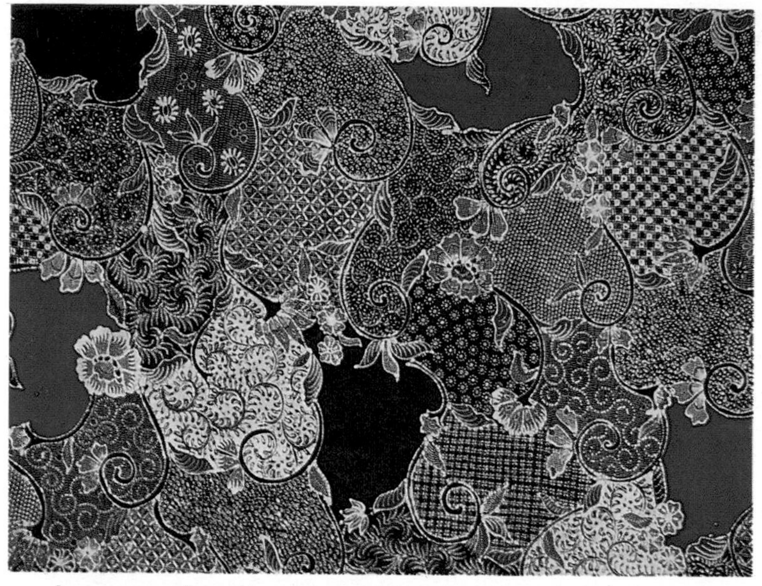

29. Isen patterns enlivened by splashes of black and red, Kenongo Workshop, Sidoarjo

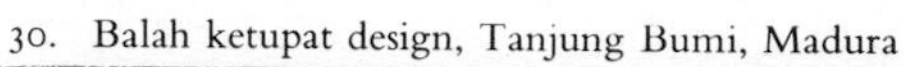

30. Balah ketupat design, Tanjung Bumi, Madura

the inner angles of the undulating scalloped edges. One of the most well-known of these interior designs is in the form of a large stylized triangular leaf which terminates in a curl. It is thought to resemble the lotus, a plant rich in symbolism in Asian art. Some Indonesian scholars regard this particular motif as a symbol of young life. On some parang examples the leaf motifs are clearly recognizable. On others (such as *parang baris*: Colour Plate 13), only the outer limits of the parang and leaf curl are portrayed, so that the design appears in the form of rows of prongs.

The variations in size, shape, and interior design of the motif give rise to different names for the parang design. A number of parang patterns feature curling tendrils and hooks, and have names such as *parang sobrah*, or 'hanging hair design'. Some parang are named after Indonesian flowers (*parang kembang*, *parang godosuli*).

Textures within the parang may also vary. Some may be cleverly crackled in such a way as to give a slightly intestinal look to the parang, while others may be covered with capillaries or dots to give a bubbly, marbled appearance to the surface of the parang. *Parang ular* (snake pattern) makes excellent use of dots and stripes to mimic the scaly skin of a reptile.

On some patterns the outer parang design has become more abstract and resembles a diagonal zigzag as in the *parang curigo* design or an eyelet pattern as in the *parang seling*. On some of the more modern patterns all that remains of the parang is a barely recognizable undulating diagonal surface design.

In addition to the mlinjon lozenge, the parang design may alternate with motifs such as the kawung and nitik and with floral patterns. On batik from Pekalongan, parang designs occasionally form the background for a spray of flowers or a fantastic bird. The parang pattern may also appear inside a circle, a diamond, or a square as in the case of *parang kurung*.

UDAN LIRIS OR 'LIGHT RAIN' DESIGN

One important group of diagonal running designs is called *udan liris*, or 'light rain' (Figure 13; Colour Plate 14). It consists of row upon row of narrow bands featuring linear examples of most of the well-known classical batik designs. Rendered in brown on a white ground, udan liris is virtually a mini-register of well-known Indonesian textile motifs. When drawn in brown with touches of blue in the background, the design may be called *rujak senti*. As with other traditional patterns, udan liris can be used to form a background for other motifs.

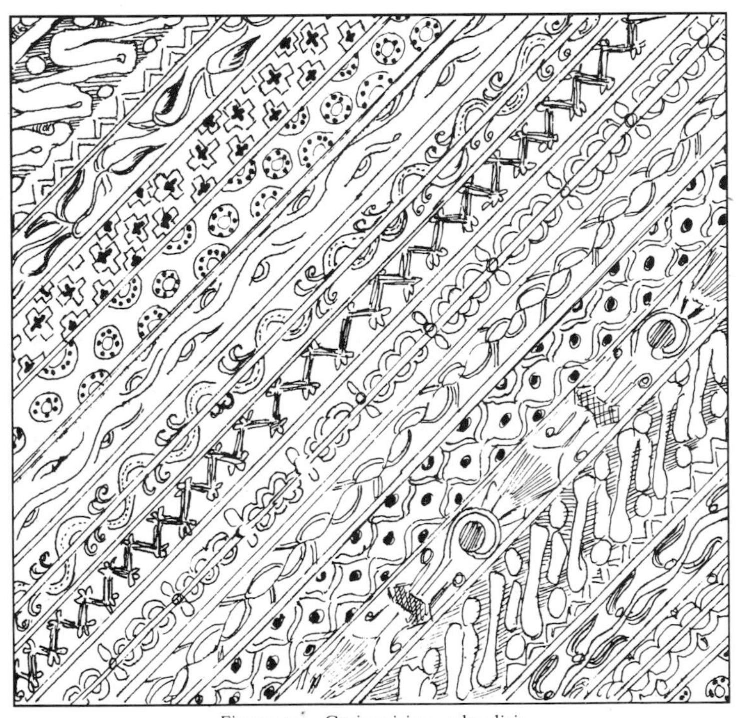

Figure 13. Garis miring: udan liris

Tambal Miring or Patchwork Design

The *tambal miring* design consists of a montage of Central Javanese batik designs set within triangles, circles, or onion-shaped lozenges arranged in horizontal or slanting rows (Figure 14; Colour Plate 15). The varying alignment of the component design elements, coupled with the contrasts between density of pattern and colour, gives the appearance of patchwork. The word tambal, in fact, suggests patched, worn-out clothes made by using left-over material of different types. This design is thought to imitate the patchwork garments formerly worn by Javanese priests as protection against malevolent influences. There is an episode from the Hindu epic, the *Mahabharata*, in the repertoire of the wayang

Figure 14. Tambal miring

kulit (Indonesian shadow play) which tells of escapades featuring Gatokaca, the son of Bima. He is said to have worn a patchwork jacket which enabled him to fly.

Tumpal or Triangular Design

The *tumpal* border design which consists of a row of isosceles triangles is a very ancient and popular art motif in Indonesia. Tumpal have been found as borders on bronze drums, and as stucco architectural ornaments on the side of ancient temple stairways in East Java.

The tumpal was also a popular border element at the narrow ends of Indian lengths of cloth imported into Indonesia. Thus, although the tumpal has long been known in the Indonesian archipelago, its arrangement on Indonesian textiles was greatly influenced by Indian trade textile designs.

The tumpal pattern usually appears at one end of a piece of cloth in the case of a sarong and at both ends on a selendang. The triangle is usually filled with floral or fauna motifs. The zigzag tumpal may be subdivided into three or four small triangles with slightly differing elements of design and contrasting colours in each sector. A small border usually delineates the boundaries of the tumpal. The area between the opposing lines of tumpal is in a darker colour. It may be plain or may contain small scattered geometric or floral motifs. Occasionally this area is quite elaborate as in the case of old batik from Lasem.

Semen or Non-geometric Designs

Some of the most imaginative and splendidly ornamented batik designs are the *semen* designs. *Semi* means 'small buds and young leaves' which are shown as curling tendrils providing an interlocking background for stylized flora, fauna, and symbolic scenes. Motifs show a wide range of Hindu, Buddhist, and Javanese inspired designs and may include elements from both European and Chinese sources.

Flower, Fruit and Leaf Motifs

Some semen designs consist entirely of flower and leaf motifs (Figure 15). Designs focusing on creepers and vines are often referred to as *lung-lungan*. Creepers of taro, ivy and fern are all represented in semen designs. Like ceplokan and parang, some semen designs are prefixed with the word *kembang* to indicate floral motifs. The *kumudaretna* floral design symbolizes wealth in its entwining plant design and may sometimes be decorated with gold (Colour Plate 16).

Creepers may also be associated with fruit and vegetables such as the chili pepper, mango, and grapes. The unripe pomegranate design, *delima wanta*, is regarded as a symbol of promise. One important traditional design is *pisang Bali*, a complicated motif showing stylized banana leaves and flowers. Palm leaves are represented by *sembagen*, a stylized fruit and flower design. The *kirna monda*, which symbolizes incomparable excellence, is represented by long leaves, tendrils, and fruit. Modern semen patterns feature flowers like the hibiscus.

Bird Motifs

Some of the most interesting of the semen patterns consist of floral designs combined with animal and bird motifs from both the real and imaginary worlds. In Chinese inspired batik, these motifs are referred to as *lokcan* and are generally depicted in lacy patterns on a light ground.

Judging from the prevalence of birds depicted on batik, the Indonesians obviously delight in feathered creatures. In Indonesian art, bird motifs have always played an important role in both symbol and ornament. Among the numerous

Lung klewer — Delima wanta — Pisang Bali — Sembagen — Kirna monda — Hibiscus

Figure 15. Semen: plant motifs

birds of Indonesia, pride of place is occupied by the garuda (Figure 16; Colour Plate 17). Originating in Hindu mythology as a half-man, half-eagle serving as the mount of Vishnu, the Preserver, in the triumvirate of Hindu gods, this bird is highly venerated and is associated with power and success. The garuda has become the national symbol for Indonesia.

Figure 16. Semen: garuda motifs

It may be seen on government buildings and on official stationery. It is also the name of Indonesia's national airline.

In batik the garuda may appear as an entire bird or, more commonly, in its feathered parts. Wing motifs, which are almost synonymous with Indonesian batik, derive from the garuda. A single wing is called a *lar*. A pair of wings is referred to as *mirong*. Two wings backed by a wide-spreading tail is a *sawat*, which is generally regarded as a symbol of prosperity and success. Wing motifs vary greatly in size and complexity. Occasionally a wing might be transformed into a creature by adding a head. The sawat ornament appears against a wide variety of backgrounds. When surrounded by small mountains and placed at regular intervals across a cloth it is called the *cuwiri* design.

Rivalling the garuda in popularity on batik is the phoenix, a motif imported from China where it has traditionally been regarded as an emblem of beauty (Figure 17). It is believed to be a composite of the virtues associated with various birds and is thought to appear in times of peace and prosperity. The phoenix may be recognized on batik by its long wavy tail feathers. The wings are usually outspread in the flying position, and the head is crested. Depictions of this bird can vary from the very naturalistic to an almost abstract swirl of wings and a tail.

The peacock, popular both in Indian and Chinese mythology, also appears in Indonesian art. It is very similar to the phoenix when depicted on batik. The rooster or chicken, traditionally a symbol of the sun, courage, and fertility appears on batik in the *ayam puger*, or 'clucking hen' design. Other birds which appear on batik are the nightingale, the pigeon, the parrot, the crow, the pea fowl, and the owl. Portrayal of these birds is very generalized and individual species are not always readily recognizable.

Phoenix

Peacock

Rooster

Ayam puger

Figure 17. Semen: bird motifs

Elephant

Deer

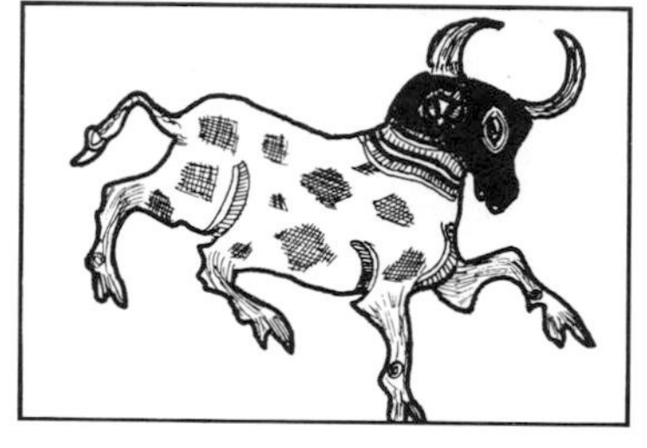

Bull

Kuda lumping

Figure 18. Semen: animal motifs

Animal Motifs

Animals may also be featured in semen designs (Figure 18). In early Indonesian art, the elephant (a symbol of strength, sagacity and prudence), the buffalo (Spring and agriculture), and the *kuda lumping* dancing horses (speed and perseverance) all appeared as mounts for the deceased. They occasionally appear in semen designs, as do tigers, monkeys, and deer.

Although not indigenous to the archipelago, the lion is a very prevalent motif in Indonesian art (Figure 19; Colour Plate 18). It entered the artistic register via Hindu culture and is often seen in wood carving, stucco, and metalwork. In batik, the Indian lion may appear on a border as a *kala* mask showing only the face. Usually, however, Indonesians prefer the Chinese lion which is depicted as a lively prancing figure with a curling mane and flowing tail. The Dutch crowned

Kala mask

Chinese lion

Chi-lin

Figure 19. Semen: animal motifs

lion has also appeared on batik. The Chinese unicorn (the *chi-lin*), a symbol of longevity, happiness, illustrious progeny, and wise administration, may also appear on Indonesian batik.

The dragon, snake or *naga* (Figure 20) symbolizing the female element, fertility, water, and the underworld is a very prevalent motif in Indonesian art. It appears on temples, on metalwork and in wood carving. On batik, naga are sometimes depicted in pairs and may be either facing or looking away from each other. They are often shown guarding things and may be depicted with huge crested heads and grinning faces. Some have wings in the shape of a lar ornament (*naga pertolo*); others are long and sinuous, while still others resemble the Chinese dragon. The naga is thought to bring luck to the wearer.

Other reptiles appearing in semen designs include the tortoise and the lizard. Fish may be seen in semen designs or

Naga

Naga pertolo

A quartet of naga guarding an object

Figure 20. Semen: animal motifs

as a repetitive element in a ceplokan design. Crustacea such as shrimps and lobsters, sea horses, jellyfish, and other marine and aquatic creatures called *bibis* or *bekingking* (Figure 21) also appear in batik design, as does the insect world. Caterpillars, scorpions, centipedes, and beetles (Figure 22) appear surreptitiously in semen designs. Butterflies and bees are popular motifs in northern Javanese batik and are usually portrayed naturalistically. A butterfly motif may sometimes fill the right angle of a border pattern.

Rock and Cloud Designs

Natural phenomena such as rocks and clouds are rendered in a most fanciful and imaginative way. Rock designs such as *wadas grompol* and *pagar wesi* and the *megamendung* cloud designs show a distinct Chinese influence (Figure 23). They are

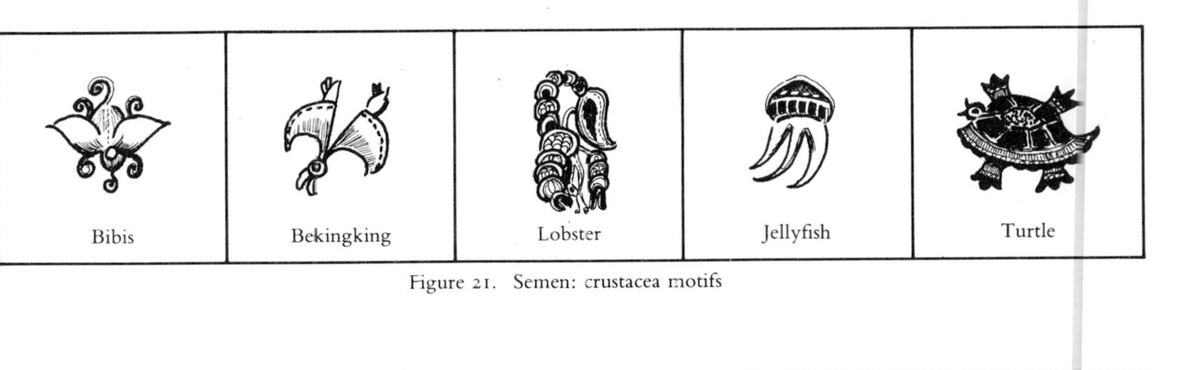

Figure 21. Semen: crustacea motifs

Figure 22. Semen: insect motifs

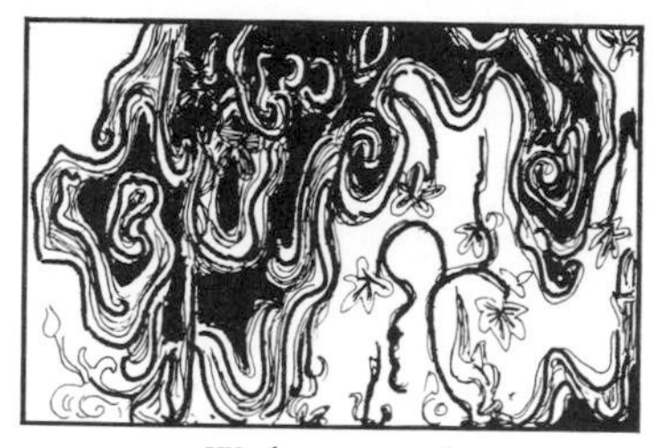

Wadas grompol

Megamendung

Figure 23. Semen: rock and cloud designs

remarkable for their layered effect. Rocks may be identified by holes and small plants sprouting from them. On the cloud motif the lozenge-shaped spirals are thicker and are usually aligned diagonally across the cloth. The inner portions of the cloud may be outlined in black. Rocks and clouds are usually portrayed in blue with scalloped or slightly undulating borders in white against a red, brown, or white ground. In Taoism, rocks form the skeleton of the earth and symbolize the creative force of nature, its strength, endurance, and majesty. The depiction of rocks and clouds in Javanese mythology symbolizes a union of the earth and sky suggestive of procreative powers.

Mountain and Landscape Designs

Mountains in batik are usually portrayed as a series of loosely connected, undulating scallops. They are often rendered in white against a darker ground and culminate in a summit which (possibly) represents Mount Meru, the centre of the Buddhist universe. In Javanese mythology mountains serve as a home for the gods, ancestors, and supernatural beings and as a place for meditation and for acquiring magical powers. Series of mountains in semen patterns subdivide the cloth into repetitive design areas depicting scenes of great symbolic and visual complexity (Figure 24).

Figure 24. Semen: mountain and landscape designs

In some symbolic mountain scenes there is a central stylized plant form complete with roots, a central stem, and branches which terminate in blossoms, buds and leaves. On some, the plant is represented by a full-grown tree, on others by a large stylized bud and leaves. A plant form may be doubled with one plant springing from the other. A plant may also be repeated as in a reflection. There are many theories to explain the symbolism of the plant form. It could represent the tree of life, the pillar of the universe from which all forms of existence spring. Alternatively, it could represent the lotus, a Buddhist symbol of purity and perfection unsullied by the mud from whence it springs, or the banyan tree under which the Lord Buddha attained Enlightenment.

Flanking the plant one often finds wings (lar) which may be strictly vegetal or may take the form of a bird with head and body. Sometimes instead of a plant, there is a small temple, or *candi*. This is a small pavilion which has a square base and was in all probability, once used for worship. Peacocks, roosters, and garuda are often seen amongst the background tendrils. Occasionally a pair of deer are seen flanking the summit of a mountain. A pair of naga may guard the temple. Density, scale, and arrangement of the component motifs vary greatly from batik to batik.

Another very popular semen design done in the same vein is the *alas-alasan*, or 'forest' design (Colour Plate 19). Within a network of tendrils backed by mountains appear a multitude of animals from both the natural and mythical worlds. It is often difficult to distinguish between the semen and animal motifs. This design in a more simple form is sometimes highlighted by gold leaf decoration.

From the Cirebon district comes the 'fragrant garden' motif, or *taman arum*. This design depicts a stylized landscape with rock grottoes, gate portals, open pavilions, fantasy animals, trees, and plants enclosed within a series of undulating cloud collars and hillocks. The motif is usually depicted on an ivory ground and owes much of its inspiration to Chinese influences. The semen tendril background is not as dense as in some other patterns.

Ship Motifs

Ship motifs, rich in symbolism, have been very important in Sumatran textiles. In batik the most popular ship motif is the *kapal kandas*, or 'beached ship' design (Figure 25). In this

Kapal kandas

European ships

Figure 25. Semen: ship motifs

design, a pair of facing birds appear on what looks like an island or undulating vertical lines. On Indramayu batik the stranded ship is quite abstract and is surrounded by what could be rocks, sand, or floating marine vegetation.

Realistic portrayals of sailing ships, paddle ships and cruisers have all appeared on batik from Pekalongan.

Human Figures

Despite a general prohibition against the depiction of human figures, Javanese shadow puppet figures or wayang kulit showing characters from Indian epics such as the Ramayana and Mahabharata, and from East Javanese legends may be seen on batik. With the increasing tourist trade, they are becoming more popular than ever. Dewi Sri, the goddess of cultivation, fertility, and prosperity (often referred to as the 'rice doll figure'), may also be seen on some newer batik. She appears as a female figure engulfed by a huge semicircular head-dress. She may appear in conjunction with plant motifs and a vase of plenty.

Scenes of Indonesian village life are occasionally seen on batik. People at work in the fields bordered by houses on stilts, children at play, officials on horseback, humble coolies, and sword-bearing soldiers have all been portrayed on batik. More turbulent times have also received attention. The writer has seen batik featuring war with tanks, parachuting air men, and heavily armed frigates. Fantasy stories such as 'Little Red Riding Hood' as well as angels and cherubs have all been depicted on batik.

Proscribed Batik Designs

To distinguish the differences in rank between various members of the sultans' families and high officials, ordinances

10. Sultan Hamengku Buwana IX wearing a kain panjang of parang rusak design with large garuda sawat. Photo copyright Museum Kraton Jogjakarta

were passed periodically at the courts of Jogjakarta and Surakarta listing batik patterns which could be worn only by members of certain status and relationship to the sultan. Commoners were expressly forbidden to wear these designs. By decrees of 1769, 1784, and 1790, the Sultan of Surakarta reserved for himself and his close relatives the following designs: parang rusak, sawat garuda wings, udan liris, and the cemukiran leaf pattern bordering the plain centre field of a head or breast cloth.

The Sultan of Jogjakarta also had certain patterns reserved for the exclusive use of the court. The parang rusak design, the sembagen pattern, and large garuda motifs were the exclusive property of the ruler, the Crown Prince, and their respective consorts (Plate 10). The other offspring of the ruler and members of the royal family bearing the title *pangeran* were allowed to wear semen patterns with garuda wings. More distant relatives bearing the title of *raden* were permitted to wear all semen designs without the wing motif. Kawung patterns and the rujak senti pattern were also permitted.

5
Centres of Batik Production

Central Java

JOGJAKARTA and Surakarta in Central Java are generally regarded as the twin capitals of classical batik, which used only three colours–blue, brown, and white–to produce the motifs previously described. These cities which were important during the Indianized period of Indonesia's history (AD 600–1500) lie at the cultural heart of Java. The remains of prehistoric man dating back half a million years have been found to the north of Surakarta. The 1,100-year-old Buddhist temple of Borobudur and the tenth-century Sivaite temples of Prambanan lie between Jogjakarta and Surakarta, while the temple-studded Dieng plateau is situated to the north. Despite conversion to Islam during the fifteenth and sixteenth centuries, the Sultanates of Jogjakarta and Surakarta have remained enclaves of ancient Hindu-Javanese culture. This influence is reflected in the classical batik design of the region.

Jogjakarta

Jogjakarta, with a population of half a million, is a centre of higher learning. It has numerous classical music and dance schools, folk theatres, and wayang puppet troupes. Many of Indonesia's leading artists, sculptors, and silversmiths live in its environs. The city is dominated by the high whitewashed walls of the *kraton* (sultan's palace) in the south-east. There are over 600 batik workshops in the Jogjakarta area.

The Taman Sari area consists of a maze of winding alleys within the palace precincts. Here are many tiny batik workshops, the owners of which proudly boast that their forebears made batik for their royal masters. Each little concern has its traditional speciality in batik design. Experts at the local textile institute are able to recognize the workmanship of some of the better-known artisans. Also in this area are a few old ladies who were formerly employees at the palace. These ladies still do canting work for special assignments which require a knowledge of the traditional reserved patterns. Much piecework, particularly the first waxing, is also done by women in outlying villages such as Buntul and Imogiri. On completion, the cloth is sent to workshops around the kraton for dyeing and finishing.

A little further to the south of the kraton in the Jalan Tirtodipuran area are approximately twenty-five family batik concerns. These factories tend to be a little larger than those in the Taman Sari area, and their batik is somewhat less traditional. The majority are family businesses employing between ten and sixty people. In all the factories that the writer visited there was a division of labour. A typical workshop such as Winotosastro which employs some sixty workers, divides the labour as follows: two women design tracers, twenty-eight women canting workers, twelve male cap workers, five men for dyeing, two women for scraping off wax, two women for painting dyes on designs, two men for boiling off the wax, seven women for sewing batik material into Western style clothing, and a further eight people for work in the showroom.

Jogjakarta designs, by and large, are very traditional and until recently, had a tendency to be static. The kawung, parang, and semen designs featuring garuda wings are firm favourites. Jogjakarta work is characterized by bold, forceful designs in dark blue and chocolate brown soga against a white background. Jogjakarta has traditionally prided itself on being the preserver of royal kraton designs which are not made elsewhere. This is gradually changing as copies of some north coast batik designs are now being made in Jogjakarta.

The city is also a centre for batik painting, an art which has become very popular during recent years. Apart from the usual tourist-trade village scenes, some very original work is being done by artists who use traditional motifs and designs to create interesting abstract pictures. Foremost amongst these artists is Amri Yahaya, a very talented young Sumatran who produces dynamic abstracts in bold bright colours (Colour Plate 20). Another well-known artist is Astuti who is noted for his Daliesque surrealist murals, while another colleague, Kuswadji, uses fine dots and marbling to give added depth and interest to his traditional and nature inspired paintings.

Surakarta

Similar in size to Jogjakarta is the city of Surakarta which is often called Solo. It is possibly the oldest city in Java. With respect to Javanese culture, it has as much to offer as Jogjakarta. It is perhaps best described as 'a Jogja without the tourists'.

Surakarta is the home of a flourishing batik industry which is generally regarded as more progressive and responsive to changing conditions than Jogjakarta's. The batik industry of Surakarta is organized on a larger scale than that of Jogjakarta. Some of the biggest batik companies in Indonesia, such as Danarhadi, Batik Semar, and Batik Keris, have their headquarters in Surakarta. Batik Keris, which has retail outlets in most of the major towns in Java, is situated on a large site where a modern complex employs 8,000 workers. A sister company, Dan Liris, produces the fabric used for batik. While this factory caters mainly to the domestic market, ten per cent of its production is for export to other ASEAN countries (Thailand, Malaysia, Singapore, and the Philippines), Japan, Europe, and Australia. Batik Keris' output runs the entire gamut of batik products from traditional canting and cap work to machine-printed fabric with batik designs. Both traditional and modern motifs are used. Traditional 'Solo batik' is famous for its natural soft brown soga dyes against a mellow yellowish background. Motifs such as the parang, kawung, and garuda wing designs are identical to those of Jogjakarta. The background isen patterns such as dots and the ukel hook pattern, are generally finer than those of Jogjakarta. The sido-mukti design of Solo is a ceplokan, while in Jogjakarta it is executed in semen style. The truntum pattern also differs on Solo and Jogjakarta batiks. The *wahyutumurun* pattern, a famous semen design featuring blue flowers and birds on a brown ukel ground, is unique to Solo. Another type of batik unique to this city is *Solo malam* (Colour Plate 21). This design consists of animal and floral semen designs in red, pink, light brown, and white against a dark brownish-black ground.

Like Jogjakarta, Surakarta still has numerous small family concerns producing traditional batik designs. In many cases the first waxing is done by women in outlying villages such as Wonogiri and Byat. On completion, their work is sent to the large factories for dyeing and finishing. Such pieces usually cost more than those completely produced within the factories because the tulis work is much finer. A village woman can spend many months waxing a single piece of cloth.

South-west Java

The batik industry of South-west Java is an offshoot of the Jogjakarta-Surakarta tradition. The towns of Garut, Tasikmalaya, Ciamus, and Banyumas are still producing batik.

Tasikmalaya

About 120 kilometres south-east of Bandung on the road to Jogjakarta lies Tasikmalaya, a city of 150,000 people. In addition to being a service centre for the surrounding countryside, this town is an important craft centre for rattan and pandanus woven goods, embroidery, and batik.

Traditional batik from Tasikmalaya is noted for its bright reds, dark greenish blues, blacks, and chocolate browns enlivened by touches of creamy yellow. Motifs include distinctive parang, stylized plants, flowers, animals, and garuda wings. On traditional brown and blue work many motifs are covered with tiny dots for decorative effect.

Today there are about twenty small family concerns on the outskirts of the town. In many factories, tulis batik using the canting is done side by side with cap and screen-printing of batik designs. Some batik is a combination of cap and canting. The outlines of the main motifs may be stamped with the cap while the details are finished with the canting.

Some of the most creative modern work in Tasikmalaya comes from the workshop of Obay Hobart who specializes in flamboyant tulis work shirt lengths (Colour Plate 22). Members of the Hobart family sketch out original designs on waxed tracing paper. The designs are transferred to 1.40 metre shirt lengths by placing carbon paper on top of the material and retracing the pattern with a ball pen. The design is arranged with a shirt pattern in mind. There are special design areas for the back, front, sleeves, collar, and pocket. These lengths of cloth, along with the wax and dyes, are given to women to do as piecework at home. The motifs are very fanciful and intricate, being composed mainly of dots, circles, and parallel lines. It takes about two weeks for a worker to complete the tulis work on a shirt length.

Garut and Ciamus

Batik from these two centres is very similar to that of Tasikmalaya. Garut, formerly known for its fine workmanship, now has only one factory practising the art of batik. A few small family concerns in Ciamus still produce both canting and cap batik. Typical motifs from Ciamus are garis miring designs, semen composed of flower and leaf ornaments connected to single garuda wings, and cuwiri patterns. Barely-recognizable birds with triple tail feathers against tritik or cecek backgrounds are also popular.

Banyumas

Batik from Banyumas has a tradition of high quality and is characterized by light reddish-brown sogas and golden yellows against bluish-black backgrounds. This town is noted for its parang designs. It is credited with inventing the famous ayam puger design and some semen floral patterns. At least two small factories are still in production, and they continue to use traditional soga dyes for special work.

North Coast Batik

The north coast region of Java extending from Anyer in the west to Banyuwangi in the east, has a batik tradition which is quite distinct from that of Central Java. Batik-making centres include Jakarta, Indramayu, Cirebon, Tegal, Pekalongan, Semarang, Kudus, Juaňa, Lasem, Gresik, Sidoarjo, Mojokerto and the island of Madura.

With the growth of commerce and trade brought about first by the Age of Discovery in the fifteenth and sixteenth centuries and later by the Industrial Revolution and colonialism in the eighteenth and nineteenth centuries, flourishing trading ports sprang up amidst the fishing villages and rice-fields of northern Java. In addition to enterprising local Javanese, Chinese, Arabs, Europeans and Eurasians came to seek their fortunes in these free-wheeling coastal cities.

With its more polyglot population, the north coast approach to batik has been much more enterprising and entrepreneurial compared with that of Central Java. The north coast region has experimented with different patterns, techniques, and textures. It was this region that pioneered the use of chemical dyes from western Europe. The use of these dyes simplified the dyeing process and greatly extended the range of colours. To meet the increasing demand for batik in the late nineteenth century, the use of the cap became widespread in the north. The technique of colet painting of small areas with minor colours, also became standard practice in the north to reduce the number of waxings. It was also in this area that batik workers who had previously worked at home, were gathered under one roof and set up as a cottage industry in the owner's compound. This development marked the beginnings of factory-made batik.

Unfettered by sumptuary laws regarding dress, and coming from different cultural traditions, the various races of the north coast all made unique contributions to the region's batik industry. The Chinese contribution to batik design may be seen in the flower and bird motifs, banji and border patterns, and in the use of the pinks, yellows and blues characteristic of northern batik (Colour Plate 23). The Arab population has generally favoured jelamprang and ceplokan patterns based on patola designs from India. They have also shared a preference for designs featuring green, a sacred colour in the Muslim world. Green has often been combined with brown, black, and occasionally purple and creamish yellow to produce a very pleasing subdued batik. A group of Dutch and Eurasian women (*Indische*) favoured European floral bouquets, complete with birds, butterflies, and bees on their batik designs. Some were known to have members of their households cut out pictures of flowers from Dutch magazines in the evenings and have their batik workers copy them the following morning. While these women were sometimes criticized for their choice of colour and motifs, their insistence on meticulous craftsmanship set standards for tulis batik which have never been surpassed. The Javanese in the northern batik industry have modified and added many traditional Central Javanese motifs in bright colours to the north coast batik repertoire.

The northern approach to batik has been essentially commercial. Profits rather than an altruistic desire to preserve a time-hallowed art have been the industry's primary motivation. Tradition and the significance of motif have been of secondary importance. Over the years, north coast batik has been geared to the varying tastes of potential customers and has reflected changes in Indonesian society. In addition to serving a considerable domestic market, north coast batik has traditionally been exported to the outer islands of Indonesia, Malaysia, Thailand and the southern Philippines where bright colours and floral motifs against detailed isen backgrounds have generally had greater appeal than the sombre Central Javanese batik.

Pekalongan

Pekalongan, half-way between Cirebon and Semarang, lies at the heart of the northern batik industry. Its batik entrepreneurs have long had a reputation for being the most innovative and most aggressive in Java. It was this city that pioneered the use of chemical dyes and the commercial use of the cap. Noted Indische women such as Mrs Eliza van Zuylen had their workshops there. It was here that the famous brightly coloured Hokukai batik was produced during the Second World War (Colour Plate 24).

While traditional Pekalongan batik in pink, blue, and yellow floral designs against intricate isen backgrounds continues to be popular, just about anything that gives itself the name batik is made here. Batik featuring both modern and traditional designs using the canting, cap, and machine printing are produced on fabrics such as cotton, rayon and silk. Quality ranges from exquisite works of art on the finest primissima cloth to the cheap and shoddy. There is also a thriving export industry in clothing and furnishings to the United States, Europe and Australia. Factories vary in size from small family concerns with three or four workers to large factories having several hundred workers. A few well-known Jakarta batik concerns also have factories in Pekalongan.

Batik entrepreneurs of Pekalongan include Javanese, Arabs, Chinese and Europeans. Well-known amongst the Javanese entrepreneurs is Achmad Yahya, a third-generation batik maker who is noted for his pastel coloured batiks for export. Achman Said of Arab-Javanese descent makes brightly coloured cap batik in floral patterns on most imaginative grounds. One of the most talented and creative people in the batik industry of Pekalongan is Jane Hendramatono, a third-generation batik artist of Chinese descent. She specializes in exotic birds, animals, flowers and rocks set against an uncluttered creamish-white ground. There is a very large Chinese operated cap batik factory called Rimbun Jaya on the outskirts of Pekalongan towards Kedungwuni.

Some of the most intricate and finely executed tulis work in batik comes from the atelier of Oey Soe Tjoen in Kendungwuni. Founded in 1924, this factory of twenty-eight workers specializes in making kain panjang, kain sarong, and wall hangings featuring Western-style bouquets of flowers and birds and butterflies against delicately drawn isen backgrounds. Traditional designs such as the cuwiri and sawat may also be executed against Pekalongan isen backgrounds. A single kain panjang may take up to eight months to complete.

Beeswax, which is obtained from local hives, is mixed with *gondurukem* (eucalyptus), dammar resins and cow fat to make a special wax combination. Canting with extra fine spouts are specially made to order. There is a division of labour in the waxing process. Two women draw the designs by holding the sample up to the light and directly tracing the design with wax onto the cloth. The fabric is then given to the most experienced canting workers to do the first waxing. Less experienced workers are responsible for waxing the reverse side and for filling in large areas during the dyeing process. Nearly one-third of the workers have been with the factory since its inception, and there are a number of second-generation workers.

The dyeing process is carefully supervised by the owner who mixes the imported chemicals himself to get all the nuances of colour required. Oey Soe Tjoen's batik, in addition to its fine tulis work, is characterized by a very subtle shading of flower motifs to create a special feeling of depth. The colours, depending on the motifs, may be applied in the following order: light blue, dark blue, yellow, and lastly brown. In a floral design, brown may be omitted and the order of colours would be blue, green, pink, and then yellow. Most of Oey Soe Tjoen's work is retailed in Jakarta. Special commissions are sometimes received. All cloth decorated at this atelier is of a uniformly high standard.

Cirebon

Cirebon on the north coast of Java lies 103 kilometres north-east of Bandung. This city has been subjected to many cultural influences. It is the meeting point for the Sundanese and Javanese, the two predominant cultural groups of Java. Arab traders from India have been coming to Java since the fourth century AD. Muslim power gained a foothold in this town in 1480. The Chinese have long been influential in the economy of the city. It is against this varied cultural mosaic that a number of unique batik designs have evolved.

The Kesepuhan and Kanoman Sultanates of Cirebon each

had their own group of batik workers who decorated cloth to their rulers' whims and fancies. The Sultan of Kesepuhan was known to be a very strict Muslim, so batik made for his court did not depict animals. The court of Kanoman was more liberal, and animals, when depicted, were cleverly camouflaged to appear as rocks, flowers and trees. It is thought that the famous taman arum (fragrant garden) motif, with its rocky grottoes, trees, and animals on an ivory ground was inspired by the gardens of the Sultans of Cirebon (Colour Plate 25).

Outside the courts, Chinese inspired designs such as the megamendung cloud motif in graded blue tones (Colour Plate 26), rock designs, banji and lokcan patterns depicting flying birds, archaic animals, flowers, and foliage on a light ground are popular. Chinese lion and dragon motifs are also very prevalent on Cirebon batik. On the finest Cirebon batik the canting workmanship is extremely fine and rhythmic. In some cases the design appears as sketched with a pen rather than with a canting.

Today Cirebon's batik industry is located in the small semi-rural villages of Weru, Trusmi, Kalitengah, and Kaliwuri near Plered some seven kilometres south of Cirebon. Over 100 small family concerns averaging fewer than ten people per workshop make batik. Their output includes the finest tulis, cap, and even silk screen-printed cloth with batik designs. This area has a tradition of using men as canting workers. Historically, some of the finest examples of Cirebon batik were made completely by men.

Possibly the best known of Cirebon's batik factories is that of Ibu Masina in Weru. This factory was founded by the present owner's grandmother in 1883 and employs between fifteen and twenty-five people. Four men produce cap batik while the remainder work with the canting. This atelier specializes in the full range of Cirebon designs. Some Pekalongan floral patterns are also produced. Local village workers are responsible for the major motifs, while workers from Pekalongan do the background isen patterns. It takes about two months to produce a megamendung design, while one featuring a detailed *sawat perganten* (garuda wing-type motif) can take up to six months.

Indramayu

A fishing village to the north of Cirebon, Indramayu has a small batik industry supporting approximately twenty-five families. Batik designs bear a strong resemblance to the lokcan designs of Cirebon. Although less sophisticated, these designs have a charm of their own. Usually only two or three colours are used–black, blue, or red on a white ground (Colour Plate 30). Designs feature the sawat garuda wings, birds, and floral and vegetal motifs. Being a seaside community, marine life in the form of fish, jellyfish, sea anemones, and seaweed is often depicted in batik designs. The kapal kandas is a popular Indramayu motif. The backgrounds of many Indramayu batiks feature pin-sized dots.

Lasem

Just over 100 kilometres east of Semarang is the town of Lasem which has a population of 4,000. With its high white walls, wooden lattice gateways, and steep tile-roofed houses, it resembles a nineteenth-century southern Chinese town.

Lasem batik was once famous for its floral motifs and exquisitely wrought Chinese motifs in deep red and blue on a light ground. Today some fifty families are officially recorded as owning batik factories. These are located in the centre of town. The batik industry is in the hands of the Chinese, many of whom have been in Lasem for seven generations. Some boast of being the third or fourth generation in the batik industry.

The workshops are located in shady, covered courtyards at the back of spacious mansions. Batik factories average about fifty workers, four-fifths of whom are women who work with the canting. There is only one factory that does cap work. Both men and women may be involved in the dyeing and wax removal process. Dyeing is by dipping and rolling material through a wooden V-shaped trough. Wax is removed by trampling with the feet and by boiling. It takes about two weeks to make a piece of average quality tulis batik. In most cases, the tulis work on such pieces is not particularly well executed. To save time the border areas are sometimes waxed with a brush rather than with a canting. A really fine piece of Lasem batik may take up to a year to complete. Such work is usually done by women in outlying villages who have the time and inclination for it. Once the first waxing is completed, the cloth is returned to the factory for dyeing and subsequent rewaxing.

Typical present-day Lasem batik is often decorated in what is called the *tiga negri* pattern (Colour Plate 27). This consists of a three- or four-coloured floral bouquet executed in pagi-sore style. Red, blue, yellow, and brown are applied in that order during the dyeing process. On one half of the cloth the floral bouquet almost merges into a dark blue-black ground which is enlivened with small circular and foliated isen patterns in yellow and cream. By contrast, the light yellow ground at the opposite end of the cloth is often covered with geometric isen patterns in the form of the kawung, parang, and *nam kepang* (weave pattern) which provide a more subdued background for the main motif.

Pekalongan and Banyumas designs, as well as simple repetitive motifs such as fish or flowers done in white against a plain or lightly decorated ground, are produced in Lasem batik factories. The finished product is retailed in the Surabaya area.

Juana

Juana, a coastal town about one and a half hour's drive west of Lasem, was once an important town for producing batik designs on silk selendang shawls for export to Sumatra and Bali. Shangtung silk from China was once used, but political upheavals in that country interrupted the import of silk and production has virtually died out. Cirebon lokcan designs on a yellow ground were very popular, as were subdued ceplokan designs on cotton for the Arab market.

Tuban

Kerek and Dongkol, two villages just out of Tuban, a small town 100 kilometres north-west of Surabaya, produce a strong coarse-weave hand-woven cotton cloth. These fringed lengths of cloth are the size of a selendang and are used to tie baskets onto the back (Colour Plate 28). They are most attractively decorated using the traditional canting process. Motifs consist of Cirebon lokcan designs featuring flying phoenixes amidst spiky foliage. The background is usually marbled due to cracking during the dyeing process. There may also be numerous pin-sized dots made with the thorn of the palmetto plant.

In Dongkol village five people are involved in making batik. This village specializes in indigo and soga dyes against a cream ground. The soga is produced by natural means. At Kerek village fifteen people are involved in the craft which specializes in blue and maroon dyes. The indigo is from chemicals, while the red comes from morinda. Kerek also produces cloth which is decorated in simple geometric designs. Instead of the usual canting, twigs from the lemon tree are used to apply the wax.

Sidoarjo

There is a small but thriving batik industry in the small town of Sidoarjo which lies twenty kilometres south of Surabaya. Approximately ten small factories in the area employ between ten and fifty workers each.

One of the largest concerns is Kenongo which has been in operation since the 1880s. While much colourful abstract batik is now being produced, a few workers devote all of their time to fine tulis batik which is produced in traditional Sidoarjo designs. A worker may spend nearly two years on a really fine piece for a kain panjang. A special wax combination consisting of beeswax, paraffin, and a black wax from Kalimantan is mixed with *gondurukem* resin to produce the fine tracery effect characteristic of high-quality Sidoarjo batik. The motifs inspired by Pekalongan are composed of brightly coloured birds and flowers set against a brown or black and cream patchwork of Sidoarjo isen patterns. Designs may sometimes be enlivened with undecorated splashes of black and red (Colour Plate 29).

Madura

Madura, a 160 kilometre-long island half an hour's ride from Surabaya by ferry, is noted for its distinctive reddish-brown batik (Colour Plate 30). Most of the batik is made in the small seaside village of Tanjung Bumi which is located on the north-west coast of Madura. About 2,000 people are involved in this cottage industry. Flower and bird motifs predominate, but in some cases the birds have become so stylized that they are barely recognizable. This could be due to the fact that the island of Madura is fundamentalist in its approach to the Muslim religion compared with the rest of Java.

All waxing is done with the canting, and the quality varies

greatly. On the cheaper pieces, the motifs set against a plain-coloured ground, are lively and rhythmic: the workmanship, however, leaves something to be desired. Dyes overlap and patches of wax may be seen on the finished product. On the more expensive pieces, the tulis work retains its liveliness, and the isen patterns consisting of fern-like foliage, parallel lines, kawung patterns, and pairs of dots in cream on a reddish ground, are carefully done. Splashes of green, red, and blue on leaves, petals of flowers, and the tail feathers of birds give an abstract dimension to the design.

At Tanjung Bumi colours are applied by laying the cloth on a wooden board and applying the dye with a scrubbing brush. Both chemical and natural dyestuffs are used. Blue comes from chemicals while yellow is from saffron. The characteristic red colour of Madura batik is produced from a mixture of natural brown soga and morinda dyes. Black may come from teak bark or from chemicals. The application of dyes by the scrubbing technique is not as thorough as dipping, and the odd patch gets missed even on the best tulis work. Possibly due to the differing waxes and dye stuffs used, batik from Madura smells different from batik made elsewhere.

Sumatra

Apart from a recently developed tourist batik industry on the island of Bali, this craft is largely confined to the island of Java. There are, however, two important centres in south-east Sumatra, namely Palembang and Jambi. These centres produce a distinct style of batik which was originally reserved as formal dress costume by the local aristocracy. Indian inspired ceplokan designs similar to those of Central Java and repetitive floral patterns are the preferred motifs. These were originally printed with wax using wooden blocks rather than canting or cap.

Traditional colour combinations include turkey red with blue and black colours. The red is generally more bluish than that of Java. More recently, brown patterns on an indigo ground have become popular. Some have been ornamented with gold on one half of the cloth and along the tumpal border. The designs on Sumatran batik, although less refined and precise than those of Central Java, are bold and lively.

Conclusion

THE greatest current challenge to the Indonesian batik industry is competition from screen-printing of batik designs. In the early seventies, a group of entrepreneurs introduced screen-printing to Indonesia. By this process, a factory with ten workers can turn out 2 000 metres of cheap, printed cotton with batik designs in a day compared with fewer than 200 kain panjang that 150 workers can produce by cap and canting. Screen-printed batik designs, being printed on one side only, lack the intensity of colour of hand-made wax-resist batik. Although it cannot compare in craftsmanship with tulis work, the price of screen-printed batik is attractive to those of lesser means.

The Indonesian government is conscious of this challenge and has encouraged the preservation of the traditional batik industry. During the late President Sukarno's time, large cooperatives (Gabungan Koperasi Batik Indonesia) were set up to import cloth and supplies and to purchase and sell batik, with mixed success. Current government regulations require producers of batik to state on their labels the quality of cloth–primissima, prima, biru, or merah–and the method by which the batik was produced–tulis, cap, combinasi tulis and cap, and machine-printed.

Jakarta has a textile museum which has an impressive collection of batik. It mounts exhibitions from time to time. A very successful exhibition was held in 1980 and a comprehensive catalogue was prepared. This museum is also used by the Wastraprema Society, an organization dedicated to maintaining the purity of Indonesian batik and weaving. There is a very active Batik Research Centre (Balai Penelitian Batik

Kerajinan) in Jogjakarta which, in addition to having an excellent collection of old batik from throughout Java, offers courses in batik. This institute conducts research on subjects such as chemical dyeing and batik design. It has also produced a number of useful publications in Indonesian.

Despite dire predictions, the batik industry has been surprisingly adaptable in the past. The invention of the cap in the late 1850s helped staunch the rush towards cheap imported European textiles which caused the demise of many native textile industries in the late nineteenth century. The use of chemical dyes, colet painting, and the development of small factory workshops, coupled with abundant cheap labour and great demand, led to an upsurge of batik production in the 1920s and 1930s. Canting workers, despite their family responsibilities, have traditionally been fairly mobile and do not seem to mind relocating with changes in the batik industry. Many canting workers from Pekalongan and Tasikmalaya work in other centres such as Cirebon and Jakarta and commute home on the weekends.

Batik design has also adapted to changing conditions. For example, a shortage of cloth during the Second World War and a need to provide artisans with work, led to the development of Hokukai batik in which beautiful bright-coloured floral patterns are set against extremely intricate isen backgrounds. Japanese taste is clearly reflected in some of these batiks. A single cloth length could take up to two years to complete.

During the time of Sukarno, K. R. T. Hardjonogoro of Surakarta, one of the foremost authorities on Javanese culture was encouraged by the late president to develop a more nationalistic style of batik. By combining his background as a scholar and artist, he sought to revitalize Central Javanese batik motifs by rendering them in the brighter colours of the north coast palette with some most pleasing results.

Sukarno also had a penchant for celebrating special events

during his reign with a new batik pattern. For example, formal visits of heads of state to Indonesia sometimes resulted in commemorative batik patterns. Designs featuring the hammer and sickle were at one time made for Communist supporters of Central Java.

Today perspectives for batik are widening. It is now possible to produce batik in longer yardages than previously. In addition to fine cotton, other fabrics such as heavy-weight cottons, cross-weaves, canvas, crêpe de Chine, taffeta, voile, silk, satin and wool are used for batik. These are now being transformed into tailored Western-style clothing, bags, hats, table and bed linen, wall hangings, lamp shades and upholstery materials.

One man responsible for the continuing interest in batik is Iwan Tirta, a Yale trained lawyer, batik scholar and entrepreneur. He has made a special study of traditional batik patterns and motifs. Using a variety of fabrics in wider widths, he has created quite a range of batik designs for fashion garments, furnishings and wall hangings. He is a preserver of the art of batik and an innovator seeking new ways to use it effectively.

Iwan Tirta has a particular flair for combining traditional designs with modern colours. He is not adverse to simplifying a motif if he considers the background too crowded; conversely he may blow up a part of a particularly attractive motif for added emphasis. He shows a particular affection for the ancient geometric designs of Indonesia. So meticulous is he in preserving standards of craftsmanship, that he has the first waxing of some of his high quality Central Javanese traditional patterns done by women in Jogjakarta who are skilled at reproducing ancient royal designs. On completion, this cloth is sent to his factory in Jakarta for dyeing and rewaxing. Iwan Tirta's designs are in great demand as fashion items both in Indonesia and abroad.

President Suharto has encouraged men to wear batik shirts

as formal wear instead of Western dress. His wife Madame Tien Suharto who was raised at the Mangkunagoro Palace in Surakarta, has a particular love for the traditional soga batiks of Central Java, particularly those made by the workshop of Princess Ayu Harjowiratmo. Her patronage and support have helped sustain an interest in classical Javanese batik.

Although many of the younger generation now wear Western dress, Indonesian women of discerning taste continue to share an appreciation of high quality batik. When buying a batik, these women will lay out the cloth and carefully inspect it for flaws. The design is noted and the tulis work examined carefully. Fine quality batik is *de rigeur* for special family events and important official occasions. To such a clientele, batik is not merely a length of cloth; it is an article of adornment to be worn, admired and treasured for it embodies all that is good and worthy of preservation in Indonesia's rich cultural heritage.

Further Reading

Arensburg, Susan McMillan, *Javanese Batiks*, Yarmouth, Maine, Ca., 1982.
Elliott, Inger McCabe, *Batik: Fabled Cloth of Java*, Clarkson and Potter Inc., New York, 1984.
Gittinger, Mattiebelle, *Splendid Symbols: Textiles and Tradition in Indonesia*, The Textile Museum, Washington, DC, 1979; Reissued with additional illustrations by Oxford University Press, Singapore, 1985.
Hamzuri Drs, *Classical Batik*, Penerbit Djambatan, Jakarta, 1981.
Irwin, John and Murphy, Veronica, *Batiks*, Victoria and Albert Museum, Large picture book No. 28, Her Majesty's Stationery Office, 1969.
Kahlenberg, Mary Hunt, *Textile Traditions of Indonesia*, Los Angeles County Museum of Art, Los Angeles, 1977.
Keller Ila, *Batik, the Art and Craft*, Tuttle, Japan, 1966.
Larsen, Jack Lenor, with Alfred Buhler, Bronwen Solyom and Garret Solyom, *The Dyers' Art: Ikat, Batik and Plangi*, Van Nostrand Reinhold, New York, 1976.
Tirta Iwan, *Batik the Magic Cloth*, Hong Kong, 1974.

Index

Numbers in italics refer to Figures.